Manfred Kröhnert

A Contribution to Resource-Aware Architectures for Humanoid Robots

Karlsruhe Series on Humanoid Robotics

Edited by Prof. Dr.-Ing. Tamim Asfour

Vol. 01

A Contribution to Resource-Aware Architectures for Humanoid Robots

by
Manfred Kröhnert

KIT Scientific Publishing

Dissertation, Karlsruher Institut für Technologie (KIT)
KIT-Fakultät für Informatik, 2016

Impressum

KIT Scientific Publishing

Karlsruher Institut für Technologie (KIT)
KIT Scientific Publishing
Straße am Forum 2
D-76131 Karlsruhe

KIT Scientific Publishing is a registered trademark of Karlsruhe Institute of Technology. Reprint using the book cover is not allowed.

www.ksp.kit.edu

Print on Demand 2017 – Gedruckt auf FSC-zertifiziertem Papier

ISSN 2512-0875
ISBN 978-3-7315-0632-4
DOI 10.5445/KSP/1000065884

A Contribution to Resource-Aware Architectures for Humanoid Robots

Zur Erlangung des akademischen Grades eines

Doktors der Ingenieurwissenschaften

der KIT-Fakultät für Informatik
des Karlsruher Instituts für Technologie (KIT)

genehmigte

Dissertation

von

Dipl.-Inform. Manfred Kröhnert

aus Ettlingen

Tag der mündlichen Prüfung:	22. Juli 2016
Referent:	Prof. Dr.-Ing. Tamim Asfour
Korreferent:	Prof. Dr.-Ing. Walter Stechele

Acknowledgement

This thesis started with my employment as doctoral researcher at the Humanoids group of the Humanoids and Intelligence Systems Lab (HIS) of the Institute of Anthropomatics and Robotics (IAR), Karlsruhe Institute of Technology (KIT) and was continued during my employment at the High Performance Humanoid Technologies Lab (H^2T) of the Institute of Anthropomatics and Robotics (IAR), Karlsruhe Institute of Technology (KIT).

First of all, I would like to thank my doctoral supervisor Prof. Dr.-Ing. Tamim Asfour for providing me with the great opportunity to work in the fascinating field of humanoid robots. I want to thank Prof. Asfour for his continuous and valuable support and advice as well as his believe in this work. Prof. Asfour's visionary ideas and the focus of the Humanoids group always served as a great inspiration and will continue to do so. Furthermore, I want to extend my thanks to Prof. Dr.-Ing. Rüdiger Dillmann who sparked my interest in robotics and provided me with the possibility of an internship at his lab before even going to university. I am very grateful for getting the chance later on to work at his lab as a student and being able to start pursuing a doctoral thesis in his Humanoids group which was led by Dr. Asfour at this time. Prof. Dillmann always supported my work and provided me with valuable feedback. I also want to thank Prof. Dr.-Ing. Walter Stechele, for being my co-supervisor, for his support, and for his helpful suggestions and discussions.

The Humanoids group has always been a great place to work at, with great colleagues, the always friendly and helpful administration, as well as the well-equipped labs. I want to thank Christine Brand, Isabelle Wappler, and Diana Becker for always taking care of anything. The great team effort of the Humanoids group made it possible for our robots to be in the position they are today. Therefore, my thanks go out to all members of the humanoids group: Dr. Eren Aksoy, Michael Bechtel, Jonas Beil, Dr. Julia Boras Sol, Dr. Martin Do, Markus Grotz, Hans Haubert, Peter Kaiser, Lukas Kaul, Christian Mandery, Dr. Michael Neaga, Simon Ottenhaus, Dr. Ekaterina Ovchinnikova, Samuel Rader, David Schiebener, Ömer Terlemez, Mirko Wächter, Dr. Nikolaus Vahrenkamp, and You Zhou. Further thanks go to my former colleagues Dr. Kai Welke for the great collaboration while building our ArmarX robot framework, Dr. David Gonzalez for being a great friend and for always providing valuable input, Julian Schill and Christian Böge for introducing me to RC-planes and who were never short of a good Loriot quote, as well as Paul Holz, Dr. Stefan Ulbrich, Fabian Schültje, Sebastian Schulz, and Dr. Markus Przybylski. Not to forget all members of the cognitive cars group, especially Dr. Tobias Gindele and Dr. Sebastian Brechtel for their friendship and valuable discussions, the programming by demonstration group, and the medical group at HIS. Special thanks go to Nikolaus Vahrenkamp for his continuous support throughout the whole thesis, for his calm and balanced nature, and for always having an open ear no matter how big his current workload. I would like to thank my good friend and office-mate Mirko Wächter for the fun times extending and improving ArmarX, for our combined and continuous supervision of the PSE student seminar (Praxis der Software Entwicklung), and for the great times flying, crashing, and repairing airplanes. Furthermore, I would like to express my thanks to Peter Kaiser and Ramin Shirazi-Nejad for their friendship, the discussions we had and the great time we spent abroad together.

This work could not have been completed without the help of my students who contributed essential parts of this work. I would like to thank Tobias Haaß, who implemented the original version of the prediction mechanisms in his master thesis and Raphael Grimm, who implemented the resource-aware motion planning algorithm in his bachelor thesis. Furthermore, I would like to thank my student assistant Adil Orhan for his work on profiling and prediction in ArmarX, and my student assistants Jan Brinker, Oliver Armbruster, and Tim Bücher for their continuous work to get the invasive computing head demo into a stable state.

I would also like to thank my wonderful friends, for the times we spent together and hopefully will continue to do so in the future. Special thoughts go to Nico Heid, one of my closest friends, who was always there for me and will always be remembered.

Most importantly, I want to thank my parents Uwe and Christina and my sister Andrea for their endless help, support, and guidance and last but not least my girl-friend Mareike for always caring and bearing with me during the finishing months of my thesis.

Last but not least I would like to thank the German Research Foundation (Deutsche Forschungsgemeinschaft, DFG) for supporting the Transregional Collaborative Research Center "Invasive Computing" (SFB/TR 89) that provided the framework in which this thesis was conducted.

Karlsruhe, July 2016 *Manfred Kröhnert*

Deutsche Zusammenfassung

Roboter werden mittlerweile in großen Stückzahlen zur Automatisierung von Produktionen in Industrieanlagen eingesetzt und sind nicht mehr aus der Gesellschaft wegzudenken. Zukünftige Robotergeneration werden sich aus den Fabrikhallen herausbewegen und sich in der Nähe zu Menschen und in für Menschen zugeschnittenen Umgebungen bewegen, sei es als autonome Serviceroboter, autonom fahrende Autos oder humanoide Roboter. Um vielfältige Aufgaben bewältigen zu können, müssen diese Roboter über eine große Zahl an Fertigkeiten verfügen, die sich an dynamisch ändernde Umgebungen anpassen können. Zu diesen Fertigkeiten zählen Sprachinteraktion, Umgebungswahrnehmung, Lernfähigkeit, Objekterkennung, Bewegungsplanung, Bahnplanung als auch motorische Fähigkeiten zum Greifen und Manipulieren von Objekten.

Eine robuste und zuverlässige Handlungsausführung bei humanoiden Robotern erfordert ein perfektes Zusammenspiel dieser Vielzahl von Fähigkeiten, sowie Algorithmen, die unterschiedliche, dynamisch veränderliche und konkurrierende Anforderungen an Rechenressourcen haben. In heutigen humanoiden Robotern werden die dafür benötigten Ressourcen oft durch mehrere Rechner mit Multi-Core Prozessoren zur Verfügung gestellt. Auf Grund begrenzt zur Verfügung stehender Ressourcen, kommt es häufig zu Engpässen und instabilem Systemverhalten, z.B. wenn alle Anwendungen die für sich optimalen Ressourcen anfordern. Wünschenswert ist ein System, bei dem drohende Ressourcen-Engpässe zur Laufzeit vorhergesagt und konkurrierenden Anwendungen kontextsensitiv Ressourcen zugeordnet werden.

Im Rahmen des spekulativen Ressourcenmanagements werden in dieser Arbeit Methoden zur datengetriebenen Erstellung kontextsensitiver Ressourcenmodelle, deren Verwendung zur Vorhersage zukünftiger Roboteraktionen und assoziierter Ressourcenbelegungen sowie deren Evaluierung auf humanoiden Robotern vorgestellt. Zusätzlich werden ressourcengewahre Algorithmen vorgestellt, die mit einer, zur Laufzeit dynamisch wechselnden, Anzahl an Ressourcen umgehen können, um ein optimales Verhalten zu gewährleisten. Damit bieten diese Algorithmen mehr Möglichkeiten bei der Verteilung verfügbarer Ressourcen.

Spekulatives Ressourcenmanagement: Ressourcenmodelle und kontextsensitive Ressourcenvorhersage

Spekulatives Ressourcenmanagement beschreibt das Management von Ressourcen basierend auf der Vorhersage kontextsensitiver Auslastungen von Ressourcen durch zukünftig anstehende Roboteraktionen unter Verwendung von Ressourcenmodellen und des aktuellen Roboterkontexts. Bei Verfügbarkeit freier Ressourcen sollen Anwendungen spekulativ und damit frühzeitig ausgeführt werden. Damit können die Ergebnisse dieser Anwendungen ohne Wartezeit bei Anfragen zur Verfügung gestellt werden. Durch korrekte Vorhersagen und vorgezogene Berechnungen kann die Ausführungszeit von Roboterhandlungen entsprechend verkürzt werden. Eine Ausführung mit falsch berechneter Vorhersage unterscheidet sich in diesem Fall nicht von einer Ausführung ohne Vorhersage. Freie Ressourcen können dabei den aktiven Anwendungen so zugewiesen werden, dass keine Ressourcenengpässe auftreten.

In dieser Arbeit werden zwei wichtige Elemente des spekulativen Ressourcenmanagements vorgestellt, zum einen das datengetriebene Erstellen von Ressourcenmodellen sowie die Vorhersage kontextsensitiver Ressourcenbelegungen.

Zur Erstellung von Ressourcenmodellen wurden zunächst Mechanismen zum Profiling relevanter Parameter während der Ausführung von Roboterhandlungen realisiert. Diese Parameter umfassen interne Roboterzustände, Umweltzustände, Ausführungsdauern einzelner Roboteraktionen, auftretende Zustandsübergänge bei der Ausführung von Aktionsfolgen, aktuelle CPU Auslastung und benötigter Speicherbedarf. Hierzu wurde das Roboter-Software-Framework ArmarX erweitert, um eine Protokollierung vorherig genannter Parameter, sowohl in Simulation als auch auf realen Robotersystemen, zu ermöglichen.

Aus den durch Profiling aggregierten Daten werden Statistiken zu Ausführungsdauern einzelner Roboteraktionen mit zugehöriger CPU Auslastung und Speicherbedarf berechnet. Diese Ressourcenstatistiken werden durch Assoziation mit gespeicherten Umweltzuständen zu kontextsensitiven Ressourcenmodellen weiterverarbeitet. Zusätzlich werden kontextabhängige Wahrscheinlichkeiten von Zustandsübergängen zwischen Roboteraktionen berechnet. Aus diesen Wahrscheinlichkeiten können anschließend probabilistische Aussagen über zukünftige Roboteraktionen abgeleitet werden. Die erstellten Ressourcenmodelle stellen die Grundlage für die Vorhersage zukünftiger Roboteraktionen, deren Ressourcenbelegungen und für das spekulative Ressourcenmanagement dar. Zusätzlich enthalten diese Modelle Informationen über untere und obere Schranken von Ausführungszeiten und Ressourcenauslastung. Mit Hilfe dieser Daten besteht die Möglichkeit Rechner in Robotersystemen geeignet zu dimensionieren oder potenzielle Flaschenhälse in existierenden Systemen zu ermitteln.

Bei der Vorhersage werden zukünftig anstehende Roboteraktionen und die damit assoziierten Ressourcenbelegungen basierend auf dem aktuellen Roboterkontext bestimmt. Dieser Kontext wird zur Laufzeit durch die bereits beschriebenen Profilingmechanismen ermittelt und besteht aus Umweltzustand, Roboterzustand, CPU Auslastung und Arbeitsspeicherauslastung. Die vorhergesagten Ressourcenbelegungen dienen als Grundlage zur Erkennung von Roboterzuständen, in denen Ressourcen überbelegt sind.

Ressourcen-gewahre Algorithmen

Die Hauptcharakteristik ressourcen-gewahrer Algorithmen besteht in deren Fähigkeit sich an eine dynamisch ändernde Verfügbarkeit von Ressourcen anzupassen, um eine optimale Systemauslastung gewährleisten zu können. Dazu fordern diese Algorithmen bei Bedarf neue Ressourcen an, geben unbenutzte Ressourcen wieder frei und passen anhand aktuell zugewiesener Ressourcen interne Parameter an. Im Gegensatz dazu verwenden die parallelen Algorithmen meist eine feste Anzahl an statisch zugeordneten Ressourcen. Dies kann einerseits zur Überlastung führen, wenn zu viele solch paralleler Algorithmen gleichzeitig ausgeführt werden. Andererseits können auch statisch zugeteilte Ressourcen unbenutzt bleiben, wenn die Ressourcenanforderungen von Algorithmen zur Laufzeit stark schwanken.

In dieser Arbeit werden zwei ressourcen-gewahre Algorithmen vorgestellt. Der Erste ist ein verteilt arbeitender Bewegungsplanungsalgorithmus, der im Roboter-Software-Framework ArmarX implementiert ist. Der zweite Algorithmus berechnet eine Tiefenbildkarte (Disparity Map) aus Stereobildern und ist in einem speziellen ressourcen-gewahren Framework implementiert.

Zur Lösung komplexer Bewegungsplanungsprobleme werden häufig parallele Algorithmen eingesetzt, die eine statische Anzahl an Arbeiterprozessen einsetzen. Diese Algorithmen berücksichtigen jedoch nicht die Verfügbarkeit von Ressourcen, wodurch Ressourcen-Engpässe entstehen können. Diese Problematik wird mit Hilfe eines verteilt arbeitenden und ressourcengewahren Bewegungsplanungsalgorithmus adressiert, der dynamisch zur Laufzeit Ressourcen akquirieren und freigeben kann. Der Algorithmus startet initial mit minimalen Ressourcen. Basierend auf einer Schätzung der Problemkomplexität werden dynamisch zur Laufzeit entsprechend viele Ressourcen angefordert. Verglichen mit der statischen Verwendung aller verfügbarer Ressourcen setzt der Algorithmus die verwendeten Ressourcen 1,2- bis 2,4-fach effizienter ein, ist dabei aber um den Faktor 2 bis 3 langsamer. Vor allem bei einfacheren Bewegungsplanungsproblemen reduziert sich dadurch die Gesamtbelastung des Systems drastisch. Die Zuteilung und der Entzug von Ressourcen kann bei diesem Algorithmus mit Hilfe eines übergeordneten Ressourcen-Managers beeinflusst werden.

Der zweite ressourcen-gewahre Algorithmus ist eine parallelisierte Variante zur Berechnung der Disparity Map aus Stereobildern. Der Algorithmus besitzt die Fähigkeit die Eingabebilder in kleineren, an die Anzahl verfügbarer Ressourcen angepassten, Einheiten parallel zu verarbeiten. Im Gegensatz zu dem vorgestellten Bewegungsplaner wurde dieser Algorithmus aufbauend auf einem Betriebssystem implementiert, das spezielle ressourcen-gewahre Mechanismen und Programmierschnittstellen anbietet. Das Verfahren wurde in Simulation und auf prototypischer Hardware evaluiert. Dabei konnte gezeigt werden, dass der Einsatz ressourcen-gewahrer Funktionalitäten die Ausführungsgeschwindigkeit erhöhen kann und die Anpassung an eine dynamisch wechselnde Verfügbarkeit von Ressourcen ermöglicht.

Contents

1 Introduction

Robots have become an integral part of our daily life, albeit being mostly hidden in factories or warehouses. Researchers are currently particularly interested in autonomous cars, autonomous service robots, and humanoid robots. The new generation of robotic systems emerging from this research will be exposed to and work in environments designed by humans for humans. These environments change dynamically during interactions between robots, humans, and objects while tasks are performed.

Therefore, humanoid robots such as ARMAR-IIIa (see Figure 1.1) have to be versatile and be able to adapt to dynamic changes in the environment arising from interactions between the robot, humans and the world. The skills of humanoid robots for performing tasks in such challenging environments include but are not limited to natural language interaction, environmental perception, learning processes, object recognition, motion planning, grasp planning, object grasping and manipulation, as well as low-level control, and much more.

Figure 1.1: The humanoid robot ARMAR-IIIa while loading a dishwasher.

1.1 Problem Statement

Robust and reliable execution of robot tasks requires perfect interplay between all of the previously listed skills whose concurrent and dynamically changing resource demands are tied to the structure of the environment. Today's humanoid robots are equipped with multi-core processors providing the required computing power. However, the limited amount of computational resources must be shared by concurrently executing algorithms. Resource bottlenecks occur in humanoid robots since, most algorithms tend to statically acquire resources in a greedy manner, not taking the global system state and resource availability into account. While static resource assignment is essential for tasks such as real-time control algorithms, higher level tasks should be able to adjust to the dynamically changing availability of resources.

In three-layer robot architectures similar to the one shown in Figure 1.2, behaviors or sub-tasks executed on the lowest level typically communicate with hardware and have very specific resource demands, while the selection of behaviors to execute is performed in the mid-level layer. A desirable goal for such robot architectures is to detect or even predict imminent resource bottlenecks and to distribute resources among concurrent algorithms in a context-sensitive manner. The goal of this work is to provide building blocks for such resource-aware robot architectures. First, the concept of speculative resource management is presented together with two essential methods for data-driven generation of context-sensitive resource models and prediction of future resource utilizations. Resource model generation methods should operate on the low- and mid-level layer of the robot architecture and capture task and behavior execution. Resource prediction should be performed on the mid-level layer and provide input for speculative resource management which can then distribute resources more efficiently on the lowest level by acting on anticipated events before they happen. Second, resource-aware algorithms are presented and shown to be capable of adapting parameters to

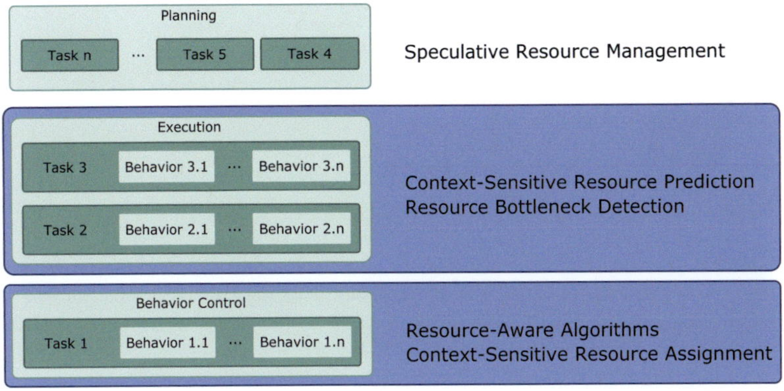

Figure 1.2: A three-layer robotic architecture (green boxes on the left) as presented in [Siciliano and Khatib, 2008]. The long-term or high-level planning layer decides on tasks required for achieving long-term goals such as preparing the breakfast in the morning, cooking lunch at noon, and setting the dinner table in the evening. Each task in a long-term plan consists of multiple behaviors or sub-tasks which can be represented as statecharts, a representation for defining the execution order and dependencies between tasks. The short-term or mid-level execution layer is responsible for selecting the appropriate behaviors for achieving a high-level task. Low-level or behavior control executes the selected behaviors by communicating with sensors and actuators of the robot. The blue boxes indicate which elements of speculative resource management influence which level of such a robot architecture.

dynamically changing resource requirements. These algorithms work on the lowest level of a robot architecture while their resource assignment can be influenced by decisions made on the mid-level, based on calculated predictions. Proposed aspects are based on resource-aware concepts and methodologies originating from the Transregional Collaborative Research Center Invasive Computing (SFB/TR 89). Their connections are depicted in a high-level overview in Figure 1.3 and explained in the next paragraphs. Invasive Computing and its main concepts are introduced in the next section.

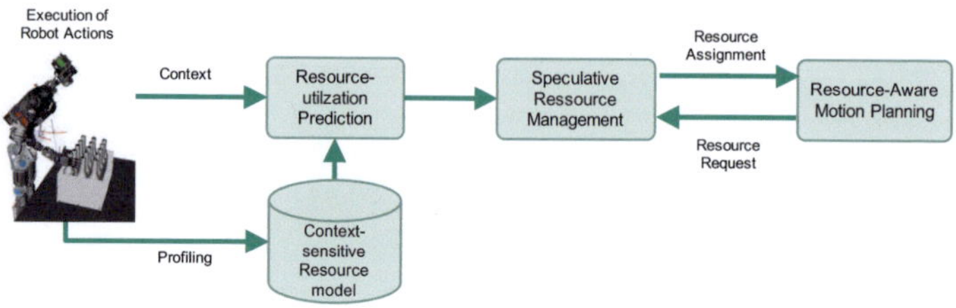

Figure 1.3: Overview of speculative resource management and the interplay between profiling, prediction, and resource-aware algorithms. Profiling methods extract data used for generating context-sensitive resource models. These resource models are further used to predict future robot tasks and their resource utilization. This prediction can be used to assign or withdraw resources from resource-aware algorithms.

Speculative Resource Management: Resource Models and Context-Senstive Resource Prediction

Speculative resource management describes the concept of managing resources through the prediction of context-sensitive resource utilization of future robot tasks, and is based on resource models and the current state of the robot and its environment. With sufficient free resources available, applications like motion planning can be started in advance. As a consequence, computations are most likely finished before results of these applications are needed, thus leading to shortened execution times. Executions with wrong predictions, however, are not distinguishable from executions without prediction. Furthermore, free resources can be assigned to running applications to increase their parallelism while preventing resource over-utilization.

The required resource models are generated using a data-driven approach which requires mechanisms for profiling relevant parameters during the execution of robot tasks. These parameters include the internal robot state, environment parameters, robot state execution times, transitions between robot states in task sequences, as well as CPU and memory utilization.

In this work, the robot software framework ArmarX was extended to enable recording of all these parameters during simulated and real robot executions.

Based on the acquired profiling data, statistics of execution times of single robot tasks and associated CPU and memory utilization are calculated. Combining these statistics with stored environment states further results in context-sensitive resource models. In addition, context-dependent probabilities for transitions between robot tasks are extracted from the profiling data.

This transition model provides the basis for determining probability distributions of future robot tasks. Furthermore, these models contain information of upper and lower bounds for both execution time and resource utilization of robot tasks. This is crucial information for speculative resource management or for determining resource bottlenecks.

In addition to resource models, this work focuses on predicting future robot tasks and upcoming context-sensitive resource utilizations. The current context required for this prediction approach consists of environment state, robot state as well as CPU and memory utilization. This context is obtained at runtime through the previously described profiling mechanisms. Predicting future resource utilization will provide the means to detect or even prevent robot states where resource over-utilization occurs in order to circumvent system slow-downs or even system failure. Additionally, the presented approach does not require specialized hardware or operating systems and runs on regular of the shelf computer systems.

Resource-Aware Algorithms

Resource-aware algorithms contain parallelizable stages and are characterized by their ability to adapt to dynamically changing resource situations. In contrast, regular parallel applications typically use a fixed amount of statically assigned resources which can lead to over-utilization in the case of multiple running applications. This behavior can result in blocking of resource, if too many algorithms run in parallel while using their optimal amount of resources. Furthermore, resource under-utilization can occur if an algorithm's internal resource requirements vary over time.

The idea behind resource-aware applications is to dynamically request and release resources at runtime and to adapt to changes in the overall resource distribution in the system. In this work, two resource-aware algorithms are presented. A distributed motion planning algorithm implemented as part of the ArmarX robot framework and a resource-aware disparity map (depth map) algorithm implemented on top of a specialized resource-aware operating system.

The presented resource-aware parallel motion planning algorithm makes use of multiple distributed worker processes. Contrary to current parallel motion planning algorithms, the amount of worker processes is not fixed and can be influenced at runtime. The algorithm starts with a minimal set of resources and requests more resources based on an estimation of the complexity of the planning problem at hand. It is also possible for an external resource manager to withdraw or withhold resources from the motion planning algorithm. This adaptive planner drastically reduces the system load, especially for simple motion planning problems, and provides the means for further load balancing through an external resource manager.

An algorithm for calculating disparity maps from stereo input images is presented as the second example of resource-aware algorithms. This algorithm is capable of processing an input image in parallel by dividing the input images into multiple smaller chunks based on the number of available processing resources. As opposed to the motion planner, this algorithm is implemented on top of the OctoPOS operating system which originates from the Transregional Collaborative Research Center Invasive Computing (SFB/TR 89). OctoPOS implements specialized resource-aware mechanisms and exposes them to application developers. Evaluation in simulation and on prototype hardware shows adaptability of the algorithm to changing resource situations and gains in execution speed when using resource-aware features.

1.2 Invasive Computing

The Transregional Collaborative Research Center Invasive Computing (SF-B/TR 89) (InvasIC) [1] aims at providing a holistic approach for programming and designing heterogeneous many-core systems containing hundreds or thousands of CPUs on a single chip [Teich et al., 2011]. Computing resources are especially limited in humanoid robots due to size and power constraints. Utilizing the computational power of compact many-core systems would therefore be a huge benefit in this domain.

InvasIC's goal is to enable the application programmer to request and release resources at runtime based on the needs of the application. The idea is that an application knows best which type and amount of resources are required at which point in time during execution. Resources considered in Invasive Computing are computational resources, memory, caches, or communication bandwidth. Realizing this kind of approach in an efficient manner requires built in support for resource-awareness in programming languages, compilers, operating systems, and hardware. All these areas are addressed by the work in InvasIC.

Once an invasive application decides to perform computations in parallel, it can dynamically request resources, adapt to the resources provided by the operating system, and afterwards release unused resources. The three phases of this request-release cycle are shown in Figure 1.4.

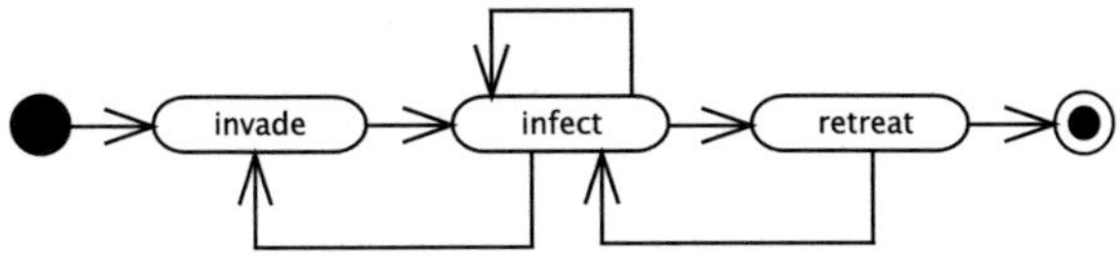

Figure 1.4: The three main resource related phases of an invasive program: `invade()`, `infect()`, `retreat()`.

[1] http://invasic.de

Initially, new resources are requested from the operating system by calling the `invade()` function which takes a collection of constraints as argument. These constraints contain information about the requested resources such as number of CPUs, amount of memory, or communication bandwidth. Furthermore, non-functional constraints can be expressed, such as exclusive or shared access to certain resources or if local instead of remote resources are requested. This `invade()` request is handled by OctoPOS [Oechslein et al., 2011], the invasive operating system. Inside OctoPOS, the agent-based Distributed Resource Management (DistRM) [Kobbe et al., 2011] is responsible for resource negotiations between applications. After negotiation is finished, a set of available resources is returned to the requesting application in a so called *claim*. After receiving a *claim*, the application has to adapt its internal algorithmic parameters to provide results with a specified quality. Adaptation is typically based on the amount and type of available resources. The application then distributes data structures onto remote and local memory resources and starts its parallel computation after issuing an `infect()` function call. `infect()` distributes isolated functions of the application to the new resources and starts them. Once resources are no longer required, they can be handed back to OctoPOS by calling the `retreat()` function, which takes the amount of resources to return as a parameter. Additionally, it is possible to `infect()` resources without calling `retreat()` and `invade()` or to `invade()` more resources after a successful `infect()`. OctoPOS exposes these resource-aware `invade()`, `infect()`, and `retreat()` operations to applications written in the C and C++ programming language. Furthermore, the parallel programming language X10 [Charles et al., 2005] was adopted by InvasIC to ease development of resource-aware parallel applications through high-level constructs already available within the language and through extensions to support Invasive Computing specific resource-aware programming constructs.

Finally, resource-aware hardware is developed in the Invasive Computing project in order to support dynamic resource allocation and to accelerate specific functions of OctoPOS in hardware. Currently InvasIC supports regular CPUs, runtime reconfigurable CPUs called *i*-Core [Henkel et al., 2011], and a specialized parallel computing hardware called Tightly-Coupled Processor Array (TCPA) [Kissler et al., 2006]. The *i*-Core is a regular CPU enhanced with reconfigurable hardware for executing special instructions. Application analyses is used to detect frequently occurring and compute intensive subroutines. These subroutines are extracted into optimized special instruction hardware blocks and a compiler annotates their occurrence in the application. These special instruction hardware blocks are then provided to the *i*-Core which dynamically loads them at runtime to accelerate the application's execution time. On the other hand, the TCPA is structured similar to Graphics Processing Units (GPUs) found in current computers and consists of a matrix of processing elements. Each processing element can be configured with a limited instruction set while the connections between neighboring elements are dynamically reconfigurable. This allows executing multiple isolated applications on a TCPA in parallel which can dynamically grow and shrink at runtime.

Within the prototype Invasive Computing hardware architecture these computational resources are grouped into so called tiles which are connected through a Network on Chip (NoC) [Heisswolf et al., 2016]. Each tile can contain additional local memory and a hardware block called C*i*C [Pujari et al., 2011]. The task of the C*i*C is to support and accelerate operating system primitives and to perform scheduling of incoming applications onto the computational resources within a tile. In addition to compute tiles, memory tiles for storing data and I/O tiles for communicating with peripheral components are supported. An exemplary hardware layout containing all mentioned components is shown in Figure 1.5.

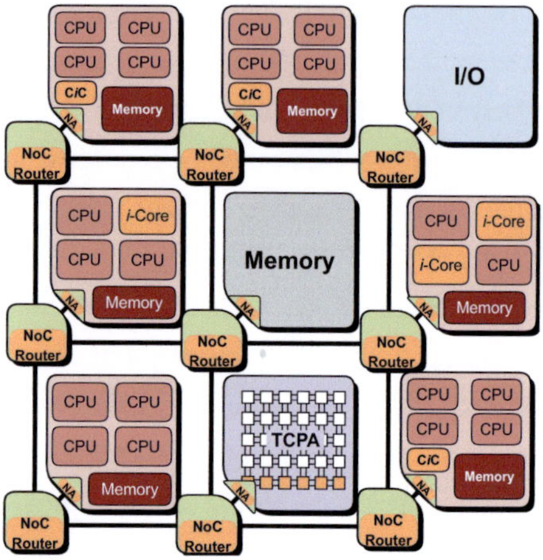

Figure 1.5: An exemplary Invasive Computing hardware architecture consisting of one I/O tile, one memory tile, 7 compute tiles, and the Network on Chip (NoC). Compute tiles are further divided into plain CPU tiles, tiles containing reconfigurable CPUs (*i*-Core), and a tile containing an array of simpler compute units called Tightly-Coupled Processor Array (TCPA). *Source:* [Paul et al., 2012] ©2012 IEEE

2 State of the Art

This chapter gives an overview of current approaches to multiple topics relevant to this thesis. It starts with humanoid robots, their architectures and the robotic middlewares used to develop software for robots. Subsequent sections cover application models, resource models as well as profiling and monitoring approaches. Furthermore, algorithms and models for prediction and algorithms for sampling-based motion planning are presented before resource-aware operating systems and similar approaches are described. All presented approaches are summarized and evaluated in the last section.

2.1 Humanoid Robot Architectures and Middlewares

Humanoid robot architectures typically consist of a hardware and software part. Hardware architectures of modern humanoid robots usually follow a distributed approach with multiple computers containing off-the-shelf CPUs and embedded components for low-level control. Specialized hardware components such as Digital-Signal-Processors (DSPs) or Field-Programmable-Gate-Arrays (FPGAs) are sometimes added for speeding up complex calculations.

Control and software architectures of humanoid robots are commonly implemented on top of a middleware layer used for communication between distributed software components.

2.1.1 Humanoid Robot Architectures

This section introduces a selection of popular humanoid robots and gives an overview of the architecture and hardware components used in these robots.

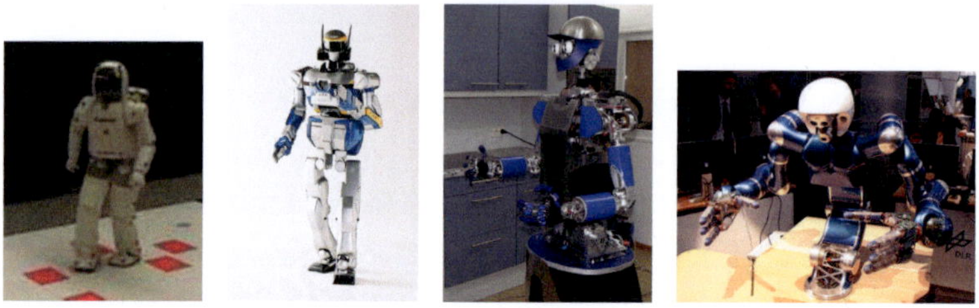

Figure 2.1: The humanoid robots ASIMO, HRP-2, ARMAR-III, and Justin. *Sources:* [Chestnutt et al., 2005] ©2005 IEEE, [Kaneko et al., 2004] ©2004 IEEE, [Asfour et al., 2006] ©2006 IEEE, [Ott et al., 2006] ©2006 IEEE

Figure 2.1 shows the humanoid robots ASIMO, HRP-2, ARMAR-III, and Justin.

ASIMO [Sakagami et al., 2002] from Honda, Japan, contains three built in PCs. One is connected to a frame grabber and performs image processing, one recognizes and synthesizes speech, and the last one executes control and planning algorithms. Furthermore, a radio communication network control unit and a DSP board for detecting sound sources are built into the robot. Additional software such as map management for navigation, task specification, or a face detection database is executed on an external server.

HRP-2 [Kaneko et al., 2004] from the Nara Institute of Science and Technology (NAIST), Japan, relies on two CPU boards using Intel Pentium III processors. Real-time control of whole body motions is performed on one board, while the sound system and other non real-time tasks such as 3D object recognition are running on the other board.

The humanoid robot ARMAR-III [Asfour et al., 2006] from the Karlsruhe Institute of Technology (KIT), Germany, distributes the workload on five embedded PCs running Linux. All PCs are connected via Gigabit Ethernet. One PC provides a gateway to the outer world and stores environment models. Three PCs are used for task coordination as well as gathering, processing, and distributing of sensor information such as camera images, laser scanner data, force torque values, audio signals and more. The last PC used for sensory-motor control is connected via CAN bus to specialized motor controllers named UCoMs [Regenstein et al., 2007] which are enhanced by DSP and FPGA components.

Justin [Ott et al., 2006] from DLR, Germany, is controlled by a PC with an Intel Dual-Pentium processor. The control software is executed on a PC running real-time QNX as operating system and is connected to a network of other PCs via Gigabit Ethernet. Non real-time tasks running on a network of PCs include computer vision, path planning, user interaction, development tools, as well as monitoring and profiling applications.

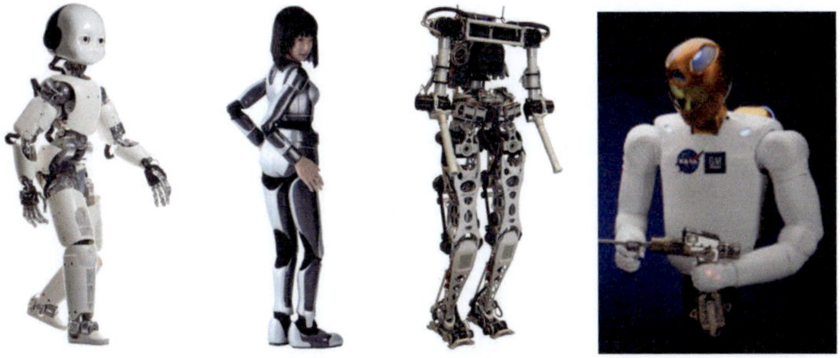

Figure 2.2: The humanoid robots iCub, HRP-4C, LOLA, and Robonaut 2. *Sources:* [Roncone et al., 2014]) ©2014 IEEE, [Kaneko et al., 2009] ©2009 IEEE, [Lohmeier et al., 2009] ©2009 IEEE, [Diftler et al., 2011] ©2011 IEEE

Figure 2.2 shows the humanoid robots iCub, HRP-4C, LOLA, and Robonaut 2.

The iCub [Metta et al., 2008] humanoid robot from Istituto Italiano di Tecnologia (IIT), Italy, contains one integrated PC104 which is connected to the outside world via Gigabit Ethernet. Motor control is realized with DSP-based controllers which rely on additional analog circuitry for communicating with the hardware.

HRP-4C [Kaneko et al., 2009], like its predecessor HRP-2, contains two built in PCs equipped with a VIA C7 and an Intel Pentium M processor. One of the PCs executes speech recognition software, while the other one performs sound control and real-time control of whole body motions which includes walking pattern generation and stabilization. Motor control is distributed throughout the robot by placing the electronics boards in proximity to the motors. Connections from controllers to the control PC are realized via CAN bus and the two PCs are connected via Ethernet.

LOLA [Lohmeier et al., 2009] from Technical University Munich (TUM), Germany, performs control on a central PC equipped with an Intel Core2Duo Mobile processor running the QNX real-time operating system. Local controllers based on custom-made DSP modules provide interfaces to motor controllers and sensors while also performing low-level tasks such as position and velocity control. The real-time capable Ethernet based Sercos-III protocol is used for communication between the control PC and local controllers. Compute intensive vision algorithms are off-loaded to an external PC cluster.

The control software of Robonaut 2 [Diftler et al., 2011] of the National Aeronautics and Space Administration (NASA), USA, runs on top of the VxWorks real-time operating system. Two PowerPC processors connected via shared memory are used to handle the workload. One processor is occupied with collecting sensor information, performing high-level sensor data processing, and overseeing the robot's safety system by evaluating kinematics and force levels in the system. The second processor performs kinematic computations and implements the control law of the robot. Joint level control is realized through electrical circuits containing embedded processors.

Figure 2.3 shows the humanoid robots PETMAN, ARMAR-4, and TORO.

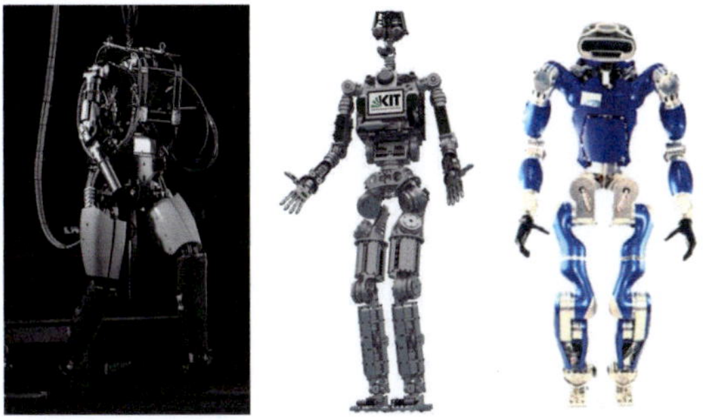

Figure 2.3: The humanoid robots PETMAN, ARMAR-4, and TORO. *Sources:* [Nelson et al., 2012][1], [Asfour et al., 2013] ©2013 IEEE, [Englsberger et al., 2014] ©2014 IEEE

PETMAN [Nelson et al., 2012] from Boston Dynamics, USA, includes a single on-board computer for reading sensors, performing high- and low-level control, logging of available sensor data, and for communication with an external operator. Internal communication uses a modified CAN bus protocol to support 1 kHz control loop frequency. The robot is controlled via a hybrid hierarchical control architecture containing a discrete component for balance control using foot placement and a continuous component for body height and orientation control and active center-of-pressure control of the feet.

ARMAR-4 [Asfour et al., 2013], the successor of ARMAR-III, contains three embedded PCs with one Intel Core i7 and two Intel Core2Duo processors which are connected via Gigabit Ethernet. Each of the three tasks perception, high-level control, and balancing are executed on one of the built in PCs. Low-level control is performed by distributed embedded components connected via CAN bus to the real-time control PC running Xenomai Linux. Off-the-shelf digital servo controllers from Elmo Motion Control [Elmo Motion Control Ltd., 2016] are used to drive all of the robot's brushless motors. These driver boards are enhanced by microcontrollers which measure absolute angles and torque and monitor temperatures.

TORO [Englsberger et al., 2014] from DLR, Germany, contains two computers equipped with Intel Core i7 processors. One of them is running a Real-Time Linux kernel for real-time control, while the other one communicates with drivers and sensors and performs high-level planning. Two more computers equipped with one Intel Core2Duo and one ARM7 processor are located in the head. They are running software for ego-motion estimation and mapping. All computers inside the robot communicate via real-time Sercos Ethernet and external control workstations are connected through wireless LAN. Communication between software components and sensors is performed via a proprietary real-time capable middleware.

2.1.2 Robotic Middlewares

Different kinds of middleware have been developed over time, each addressing different concerns of developing robotic applications. The presented middlewares are all distributed under an open source license and are designed for building distributed robot applications in order to take advantage of the increased processing power of multiple computers.

Design goals of the Orocos [Bruyninckx et al., 2003] framework include being portable and real-time capable. Portability is achieved through the implementation of an Operating System Interface (OSI). Robot programs are written as loosely coupled distributed components and communication happens via sending events. Orocos heavily relies on general software design patterns and those used in distributed systems.

The OpenHRP [Kanehiro et al., 2004] framework is used in the HRP robots (such as HRP-2 and HRP-4C) and builds on the standardized CORBA middleware to supports writing modular and distribute software components. The CORBA Interface Definition Language (IDL) is used to describe component interfaces in a standardized way. Portability is achieved by abstracting access to the robot hardware into interfaces which must be implemented for simulation and for the real robot.

YARP [Metta et al., 2006] is used in the iCub robot and is designed as a framework for building distributed and modular robot applications. Relying on the Adaptive Communication Environment (ACE) [Schmidt, 1993] for communication and operating system abstraction makes YARP portable among different operating systems. Communication is built around the Observer pattern where one or multiple observables provide data through so-called Ports which can be observed by multiple consumers. Connections between Ports can be established and removed dynamically at runtime. This makes the software tolerant to crashes, since connections can be re-established after restarting an application. Additionally, timing inside an application can be decoupled since each Observer can process incoming data at its own speed.

The Agile Robot Development toolkit (aRD) [Hirzinger and Bauml, 2006] is used in Justin and was developed to support real-time applications and to allow control-loop frequencies above 1 kHz. It builds on the basic concept of an execution block which can be executed independently with a fixed priority. Execution of a block can either be triggered periodically or through incoming data. An application is built as a distributed network of blocks and communication links. Connection schemes between blocks are static at runtime. Tools are provided to specify these links and to distribute the blocks of an application onto specific hardware components. In contrast to the other described frameworks, aRD relies on specifying the functionality of blocks in Matlab/Simulink which is later on compiled into executable programs.

The Robot Operating System (ROS) [Quigley et al., 2009] was designed to address issues in large-scale integrative robotics research where scale, scope, and variety of developed software continues to grow. It is currently one of the most widespread robotics middlewares. The four main concepts of ROS are nodes, messages, topics, and services. Nodes are software components realizing a specific functionality. These nodes communicate in a distributed peer-to-peer network via messages. It is possible to send these messages via unidirectional topics to multiple nodes which subscribe to the topic in order to get notified upon arrival of new messages. Services on the other hand provide means for synchronous and bidirectional communication and are specified by a name and a pair of request and response messages. Similar to YARP, single nodes can be restarted at runtime without interrupting the complete application network. Additionally, ROS provides a collection of tools for visualizing, monitoring, compiling and launching software modules. Contrary to YARP, which uses the ACE library for communication, the message-based format and communication infrastructure of ROS are currently non-standardized, with efforts being made to switch to a standardized messaging protocol.

The ArmarX [Welke et al., 2013, Vahrenkamp et al., 2015, Wächter et al., 2016] framework is used in ARMAR-III and ARMAR-4 and aims at providing the building blocks for creating high-level robot architectures. The major design goals are distributed processing, interoperability, robot independence, and disclosure of the system state. On the lowest level, ArmarX relies on the Internet Communication Engine (Ice) [ZeroC, Inc., 2015] which provides the infrastructure for writing distributed applications, defining interfaces with the included Interface Definition Language (IDL) Slice, cluster management, different types of communication, and other types of services. Several layers are built on top, beginning with the middleware layer which provides life-cycle and dependency management of distributed framework components, graphical specification of control- and data flow, as well as programming interfaces for controlling actuator units, for reading sensor data, and for processing sensor data. The robot framework layer provides generic robot components and interfaces, the memory layer provides biologically inspired mechanisms for storing and persisting data, and the vision layer provides interfaces for accessing and processing images, video streams, and point clouds. Hence, ArmarX enables users to create software architectures for robots by reconfiguring as many generic ArmarX components as possible while providing interfaces for easy integration of robot specific components.

Furthermore, the disclosure of the system state principle in ArmarX is supported by tools which aid in developing, visualizing, and debugging robot applications.

2.2 Application Models

This section focuses on approaches for defining high-level control flow in robot tasks. There are however other types of control flow definitions available, such as control flow graphs which are found in compilers. Those are not covered, since they focus on application details which are too low-level for this work.

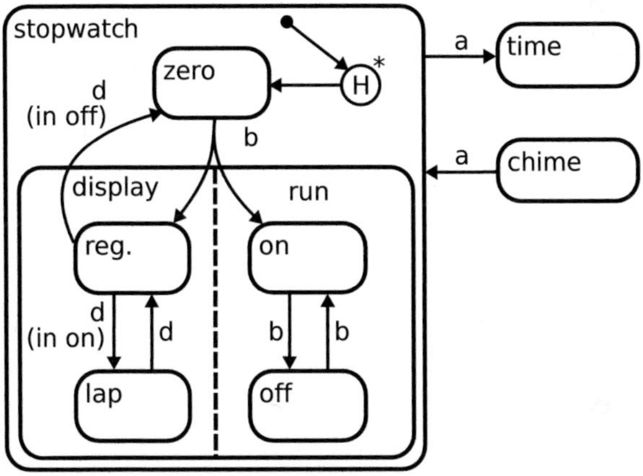

Figure 2.4: An excerpt of a more complex statechart showing the logic of a stopwatch. *Source:*[Harel, 1987]

Statecharts introduced by Harel [Harel, 1987] are an extended version of finite state machines (FSM) and provide a structured approach for representing complex control flow. Major extensions include: support for hierarchical grouping of states, parallel execution of states, history in states, and jumping directly to sub-states anywhere in the hierarchy. Those extensions are backed up by a formal definition of their semantics [Harel and Naamad, 1996]. Figure 2.4 shows a section of a more complex statechart presented in the original statechart paper. It shows a hierarchical statechart which includes a history state *H*, parallel execution of the sub statecharts *display* and *run*, and jumping between hierarchies (from the state *reg.* to the state *zero*). Overall, statecharts provide a structured way to develop the control flow of complex applications.

The Palladio Component Model (PCM) [Becker et al., 2009] is a framework for describing component-based software architectures. Through specification of functional and non-functional parameters it is possible to reason about the performance of the model when it is executed on different types of hardware. Interactions between components in the PCM are described as *Service Effect Specifications* which in turn are modeled as finite state machines. In comparison to regular FSMs, the PCM state machines are extended with transition probabilities used to predict the probability of component failures.

The Unified Modeling Language version 2 is a general-purpose modeling language from the field of software engineering [Object Management Group (OMG), 2015]. It was designed to create a common notation and visualization system for software architectures and single pieces of software. In UML 2, state machines are used to describe the behavior of software while the software structure is defined by other UML 2 mechanisms. The state machine semantics are described in chapter 14 of the current UML 2 specification (version 2.5). Overall, UML 2 state machines are described as an object oriented version of Harel's statecharts, tailored to describe behaviors in UML 2 based software architectures.

Within the context of robotics, state machines and statecharts are used for describing robot tasks and robot behavior.

The Task Description Language (TDL) uses dynamically generated task trees as its basic concept. Its description and formalization is presented in [Simmons and Apfelbaum, 1998]. Two types of nodes exist in the task tree: command nodes and goal nodes. Command nodes are used to specify runtime behavior which is executed upon entering the node. Goal nodes are

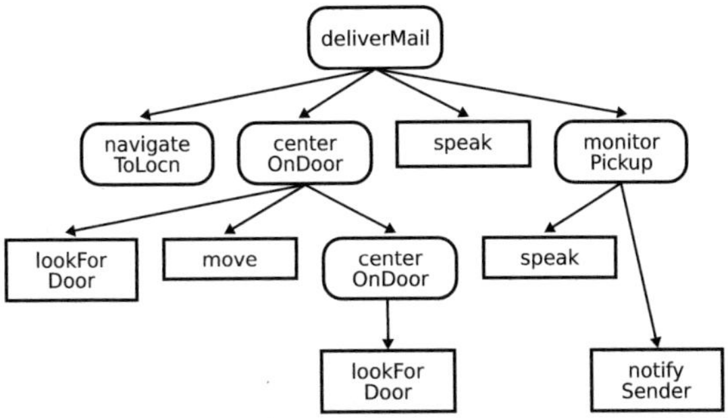

Figure 2.5: An exemplary task tree of the Task Description Language (TDL). *Source:* [Simmons and Apfelbaum, 1998]

used to represent high-level behaviors and are expanded at runtime into a tree of goal nodes and command nodes. Figure 2.5 shows an exemplary task tree with expanded nodes. Command nodes as opposed to goal nodes can only have other command nodes as children. Overall, TDL was created to formalize specification and implementation of robotic tasks. It provides means similar to activity diagrams found in UML 2 but differs in the dynamic expansion of nodes at runtime which allows for a more compact description than activity diagrams.

The event-based Task Description Language (ETDL) presented in [Park et al., 2006] was created as a task description language to be used by a task monitoring system. Design goals for the language were hierarchical task specification, expression of variations in task executions similar to regular expressions, and the ability to forecast tasks described by the language. Tasks themselves are described as a behavior associated with pre- and post-conditions. However, no qualitative comparison of the new approach and existing task description languages is presented and no formal semantic specification is given.

Extended State Machines (ESM) were introduced as an extension to classical state machines to help designing, implementing, and testing control systems of autonomous robots [Merz et al., 2006]. Additions made include task states for executing user defined program code, states with timers, data flow between states, and special flags to configure sub state machines and react on their outcome.

A model-driven approach for designing robotic software is described in [Schlegel et al., 2009, Steck and Schlegel, 2010]. This work is an example of how UML can be applied for designing software for autonomously driving robots while using statecharts to specify the behavior of the software and thus the robot.

The restricted Finite State Machine (rFSM) approach implements a minimal variant of Harel statecharts and UML 2 state machines [Klotzbücher and Bruyninckx, 2012]. Target properties of the rFSM implementation are composability of states, compositional robustness, and satisfiability of real-time constraints. The minimal subset is defined in the paper as "the smallest number of primitives necessary for humans to construct practical coordination statecharts". To achieve this minimalism, rFSMs omit parallel state execution as well as the execution time requirement, which requires that effects of an action in step n can only be sensed in step $n + 1$. The rFSM model is integrated into the real-time toolkit of the OROCOS robotics framework and is used for explicit coordination of components in a robotic software architecture.

Successful application of statecharts is also presented in [Paraschos et al., 2012], where statecharts are used to coordinate individual soccer playing robots as well as a team of those robots. Both functional and behavioral aspects of the control of a single robot and a team of robots is described via statecharts. After designing the controllers, tools are used to generate C++ code which can be compiled and run on the real robots.

A domain specific programming language based on UML/P statecharts is described in [Thomas et al., 2013]. UML/P is an implementation oriented version of the UML 2 specification, and allows for generating Java code out of specified models.

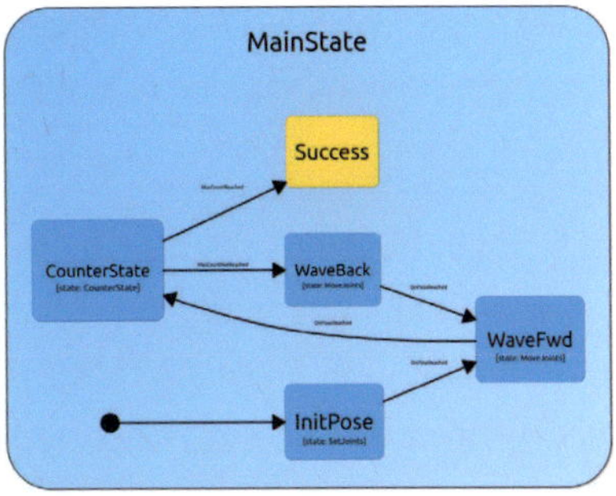

Figure 2.6: The visual representation of a statechart in ArmarX. The initial state is marked by the transition from the black dot. Regular sub-states are shown in blue and end-states are shown in yellow.

The robot framework ArmarX also contains an implementation of hierarchical statecharts [Welke et al., 2013, Wächter et al., 2016]. High-level control flow within ArmarX applications is realized via statecharts. Figure 2.6 shows the graphical representation of an examplary ArmarX statechart. In addition to classical statecharts, it is possible to specify the data flow within these statecharts. Only sensor data acquisition and hardware access is implemented as specialized components which are called from states within the statecharts. One major aspect of ArmarX statecharts is their distributed nature which allows running different parts of the complete robot program on different computers. Additionally, runtime substitution of states is possible and is used to implement high-level planning which decides on executing a specific statechart at runtime.

2.3 Resource Models

Modeling resources in order to describe resource demands and resource utilizations of applications or algorithms is a topic found in multiple domains. Most existing modeling approaches show that resource models can only be interpreted in a meaningful way if they are expressed based on a reference model of the platform the respective algorithms and applications are executed on. The following paragraphs give an overview of how resources are modeled in different domains.

High Performance Computing (HPC) deals with massively parallel algorithms executed on cluster computers with hundreds to thousands of processors. Optimal resource allocation is crucial in this field for maximizing performance and reducing operational costs. Hence, resource models are extracted from HPC applications on which resource scheduling is performed as described in [Subramoniam et al., 2002]. Resources considered in this publication are classified as CPU cycles, disk space, memory space, network bandwidth, and specialized processing power such as GPUs. However, no specific units for each class of resources and no explicit model of the execution platform is given. Thus, the resource model is tied to a specific cluster architecture and before moving to a different architecture, a mapping between the old and the new platform must be established.

Scheduling theory provides the mathematical foundations for computing a feasible schedule providing real-time guarantees for executing the workload of a number of applications on a given set of resources. Hence, this field of research is another prominent area where resource models are used. In [Shin and Lee, 2003] a Periodic Resource Model (PRM) is presented, and is expanded in [Easwaran et al., 2007] towards a generalized Explicit Deadline Periodic Resource Model (EPD). The goal of the model is to accurately describe periodic resource allocation behaviors of periodic tasks by guaranteeing the allocation of Θ time units every Π time units. Schedulability analysis performed with these models is thus determined based on time units. Resource demands of workload sets are specified as the amount of required resource allocations requested by the workload set. However, the definition of these resource models is based on the abstract notion of time units. The presented resource models do not take fluctuations in workload execution times or resource demands into account.

Based on the well established UML 2 standard, the "UML profile for Schedulability, Performance, and Time Specification" (SPT) was developed and standardized by the Object Management Group (OMG) [Object Management Group (OMG), 2005]. Since 2011, it has been replaced by the "UML Profile for MARTE" [Object Management Group (OMG), 2011]. One idea behind MARTE is to develop software based on an abstract hardware model and later on map the software to a specific hardware model determined via performance analysis. Non-Functional-Property types (NFP) such as utilization, basic measurement units, and probability distributions defined as operations on NFP types are multiple ways to describe resources using MARTE. The Generic Resource Modeling (GRM) library provides means to model resources and resource demands. These resource demands are specified as a resource amount and a reference to a resource type. Resource amounts are further specified as Behavior which can be any of repetitions, probability, hostDemand (CPU), priority, responseTime, executionTime, interOccurrenceTime, throughput, utilization, utilizationOnHost, and memory. Additionally, the *ResourceUsage* package of GRM provides means to express statistical and dynamic usage of memory, CPU time, number of Bytes sent through a communication resource and more. The "Hardware Resource Modeling" library (HRM) is a refinement of the GRM library [Taha et al., 2007] and lists the following hardware specific ResourceTypes: SynchResource, StorageResource, ConcurrencyResource, CommunicationResource, ComputingResource, TimingResource, and DeviceResource. The "Software Resource Modeling" library (SRM) refines the GRM library [Thomas et al., 2008] towards the software specific aspect and lists SchedulableResource and CommunicationResource as software specific resources.

The Palladio Component Model (PCM) from the software modeling domain focuses on the development of Component Based Architectures [Krogmann and Reussner, 2008, Becker et al., 2009]. PCM does not extend the general UML2 model like MARTE, but instead creates a new separate model focusing on modeling and analysis of performance predictions and advanced software component concepts whose constructs are mappable to the performance domain. Resource types covered by PCM are communication bandwidth, CPU, and memory which are expressed based on abstract *ResourceTypes*. For example, CPU demand is specified as time units of an abstract *ProcessingResource*. Resource demands are specified as constant quantities, probability distributions, or parameterized functions. The resource demands are part of a so called Service Effect Specification (SEFF) which contains dependencies on other components and expected input parameters or expected return values of used services. Abstract *ResourceTypes* are instantiated later in the development process in a *ResourceEnvironment* which specifies available resources, concrete resource types, and connection between resources. Performance of a software architecture is then evaluated by mapping it onto multiple *ResourceEnvironments* and running simulations. Furthermore, PCM contains concepts for expressing component Quality of Service (QoS) parameters as parametric contracts depending on the component's context [Becker et al., 2006]. This context is specified independent of the component implementation and is defined as the software environment a component instance is contained in. Parameters like deployment on concrete hardware, used system software, or access patterns directly influence the component performance and can vary from instance to instance. Finally, the PCM approach of component contexts deals with static aspects of the software architecture which are fixed or computed at design time.

A simple resource model containing CPU and memory consumption is used in [Park et al., 2012]. Each component of the robot software is annotated with an expected resource consumption which is computed through offline simulation of the component's execution. During execution, these parameters are replaced by runtime measurements of running components. These resource demands are used during task planning for generating resource constrained plans. CPU usage is measured in percent and main memory consumption in MByte. No explicit execution platform is specified which makes it difficult to transfer the results from one platform to a different one.

A model-driven approach for robot software design which makes resource requirements explicit is presented in [Steck and Schlegel, 2010]. Resource requirements are modeled as non-functional properties (NFP) and are used for design time analysis of resource constraints. These NFPs include worst-case execution time and the period of tasks which are used to perform schedulability analysis of real-time tasks. Without further detail, the paper mentions that some properties will be reasoned about at runtime since they can be unknown at design time.

2.4 Profiling and Monitoring

Resource models (Section 2.3) or transition probabilities in PCM state machines (Section 2.2) are either based on expectations or are extracted from usage data of real or simulated application executions. Different profiling and monitoring approaches for acquiring usage data from applications are presented in this section. The focus for types of data to acquire is put on CPU and memory utilization, communication, state transitions within a statechart, and environment changes.

The Task Description Language (TDL) specifies programs as a tree of command nodes and group nodes. Additionally, task decomposition, synchronization, execution monitoring, and exception handling are supported by TDL [Simmons and Apfelbaum, 1998]. Any task in the task tree can define expressions depending on monitored values. Once an expression evaluates to be true, an event is triggered and sent to the task that initially provided the expression. These events are triggered by a special Monitor task based on external events, state transitions between task tree nodes, or due to passage of time.

A Plan Execution Monitor (PEM) for failure recovery in high-level task planning is presented in [De Giacomo et al., 1998]. Execution monitoring is described as "the robot's process of observing the world for discrepancies between the actual world and its internal representation". Based on knowledge gained through execution monitoring it is possible to react on and recover from discrepancies. This work is based on high-level symbolic knowledge of the world state and all possible actions. The Execution Monitor checks if the world state matches the precondition of an executed action and triggers a recovery action in case of a mismatch.

A survey of execution monitoring methods in robotics is presented in [Pettersson, 2005]. The survey was conducted because robots must adapt to the current environment which in turn requires execution monitoring to detect faults and environment changes. Three categories of approaches are described: analytical-based fault detection, data-driven approaches, and knowledge-based approaches. Analytical-based monitoring and fault detection approaches rely on mathematical formulations for describing system or robot behaviors. The most prominent version in this category are observer-based methods which compare an estimated system state with the actually observed system state, such as the PEM in [De Giacomo et al., 1998]. Data-driven approaches do not rely on mathematical models, but instead derive information for fault detection from statistic measures computed from measured input data. Knowledge-based approaches combine both analytical-based and data-driven approaches. They are provided with training data categorized by human experts and try to decide in a similar way. The survey states that analytical observer-based monitoring approaches are commonly used in robotics due to their correlation with state machine representations of robot programs. In order to detect possible failures, observer-based approaches compare the actual system output against an estimated system output which is calculated from the last known input and output parameters.

A model-driven approach for robot software design is presented in [Steck and Schlegel, 2010]. The work discusses reasoning about non-functional properties at runtime and changing the component wiring depending on the current situation of the robot. However, the work focuses on the software modeling aspect of designing robot software and no further details on monitoring the current situation are presented in the paper.

Resource usage monitoring is used for executing resource constrained plans of robot actions as described in [Park et al., 2012]. This approach uses a framework specific Positioner component to monitor the CPU and main memory resource status of a Single Board Computer (SBC). The planning component yields a resource constrained execution sequence of component candidates. Afterwards, the Positioner is responsible for selecting the appropriate components to execute, based on the current workload of the SBC.

A model for self-monitoring during execution of robot tasks is presented in [Kurup et al., 2013]. The authors describe a model for classifying the robot's status by associating sensor information such as position and speed of a robot with the robot's internal operational state (Moving, Looping or Stuck). This model is trained during an initial learning phase and maps previously recorded sensor information onto simultaneously executed robot behaviors of a navigation task. Afterwards, the same task is performed by the robot and the learned model is used to diagnose problems during execution and to detect the progress within the navigation task. During execution, possible gaps in the learned model are filled by consulting a rule based heuristical model. Since the rules are based on low-level sensor information a lot of domain knowledge is required for manually specifying them.

The ArmarX robot framework emphasizes state disclosure as an essential requirement for implementing robot capabilities [Welke et al., 2013, Vahrenkamp et al., 2015, Wächter et al., 2016]. Within the framework, robot programs are represented as hierarchical distributed statecharts (see Section 2.2). The internal structure and all incoming and outgoing data of these statecharts are exposed for online inspection of the system state. Additionally, the robot memory can also be inspected which contains information about the state of the environment as perceived by the robot. This enables monitoring of the currently executing state, transitions in the statechart, and the environment of the robot.

Advanced ROS Network Introspection (ARNI) is an extension to the standard ROS robotics framework [Bihlmaier et al., 2016]. It allows generating statistics for hosts running ROS programs and includes communication bandwidth and CPU usage. Furthermore, constraints for the generated statistics can be specified for executing counter measures if the constraints are violated.

Several approaches exist for profiling CPU usage of applications on Linux based operating systems. They range from statistics generating compile-time enhancements, application execution on a virtual CPU, and reading CPU internal performance metrics to reading metrics provided by the Linux kernel.

Both gprof [Graham et al., 1982] and Google PerfTools [Google, 2005] fall into the category of modifying programs at compile-time and later on at runtime. The gprof profiler enhances an application at compile-time with special functions used to generate statistics at runtime. Google PerfTools are added as libraries to an application at compile-time and also generate statistics at runtime. After the execution of an application, the statistics of both profilers are written to disk and can be evaluated afterwards. In many cases, neither can be used due to required recompilation and the extra overhead for calling the special functions.

Very accurate measurements can be retrieved by profiling an application with tools of the Valgrind framework [Weidendorfer, 2008]. This approach executes an application on a virtual CPU and intercepts calls to the hardware for generating precise statistics. However, there is a notable slowdown in execution time of at least a factor of 10 in using this approach.

Modern CPUs are equipped with embedded Performance Counters which contain processor specific information such as the number of executed instructions, the number of floating point operations, or the number of cache misses. One possible way to obtain these values from within an application is to use the Performance Application Programming Interface (PAPI) [Browne et al., 2000]. System wide profilers such as OProfile [Levon et al., 2002] or perf [Kernel.org, 2010] also make use of Performance Counters by generating statistics for single applications or complete operating systems.

These statistics are written to disk after program execution and can be analyzed with tools provided by the profilers.

Both SystemTap[2] [SystemTap Documentation, 2005] and DTrace [3] [McDougall et al., 2006] are system wide profilers with support for application specific profiling. Statistics such as CPU usage are gathered based on user supplied profiling scripts.

The Tuning and Analysis Utilities (TAU) framework targets "performance instrumentation, measurement, analysis, and visualization of large-scale parallel computer systems and applications" [Shende and Malony, 2006]. It is used for gathering performance data such as CPU utilization, memory consumption, or communication bandwidth of applications running on parallel CPU architectures such as clusters.

Operating system provided metrics can be obtained by using the GNU C library (glibc) which provides interfaces for reading the CPU usage at runtime [GNU Project, 1997a]. Additionally, CPU usage of applications running on Linux based operating system can be extracted from information provided by the kernel via the `/proc` filesystem [Linux Kernel Documentation].

Dynamic memory usage profiling of an application is supported by some of the CPU profiling tools. Namely, Valgrind Massif [Valgrind Documentation, 2004], Google PerfTools [Google, 2005], TAU [Shende and Malony, 2006], SystemTap [SystemTap Documentation, 2005], and DTrace [McDougall et al., 2006].

The Linux kernel exposes memory statistics via the `/proc` filesystem, similar to CPU statistics [Linux Kernel Documentation]. However, only virtual memory statistics are provided which do not contain information about dynamic allocation and deallocation of memory.

Thus, some approaches replace the memory manager to gain access to allocation and deallocation statistics. In C++ code it is possible to override the `new()` operator within an application or to replace the calls to memory

[2] *SystemTap* only runs on Linux.
[3] *DTrace* does not run on Linux.

allocation function `malloc()`. However, both approaches require the user to do manual bookkeeping.

In general, the GNU C library provides an interface to read virtual memory statistics in combination with CPU usage via the `getrusage()` function [GNU Project, 1997a]. However, the memory fields are not supported by Linux kernels. Instead, it is possible to retrieve memory statistics of the C library's memory allocator via the `mallinfo()` function [GNU Project, 1997b].

2.5 Prediction Models and Algorithms

A wide range of prediction and machine learning algorithms exists that are applied in embedded systems, autonomous cars, and many other applications. The goal of this section is to present a selection of relevant approaches capable of predicting the state of robot applications which are based on application models presented in Section 2.2. Additionally, the prediction models should be learned from experience based on data collected with methods described in Section 2.4.

The book "Machine Learning" contains a comprehensive collection of machine learning algorithms [Russell et al., 1996]. Amongst others, it covers Bayesian Learning, Markov Chains, Hidden Markov Models (HMM), and Support Vector Machines (SVM) which can be applied for calculating prediction.

Dynamic Bayesian Networks (DBN), a variant of Baysian Learning, are shown to be suitable for learning a navigation task model in [Infantes et al., 2006]. In this work, a robot performs a navigation task where it encounters different situations such as a free path, a narrow path, or a blocked path. First the DBN based model has to be learned from previous executions of the task so that future situations can be reliably predicted by the model. The DBN model is also used to generate plans for robot actions and can be constrained by objectives such as avoiding to maneuver the robot into a failure state where it gets stuck. As expressed in [Russell et al., 1996], DBNs are a generalization of Hidden Markov Models providing a probabilistic representation in combination with inference (prediction) mechanism for time-dependent domains.

Markov Chains are used in [Albers et al., 2009] to model short term changes of task execution times in an image processing pipeline. Depending on its content, an image is processed with different combinations of algorithms within the pipeline. The image processing time varies with the execution times of the applied algorithms. Using the Markov Chain model, switches in the pipeline control flow are predicted reliably and used to calculate upcoming resource usage patterns. The Markov Chain model is used in this case since the internal state of the system is known/observable and previous execution steps in the pipeline influence future decisions on how to process the image.

Case Based Reasoning (CBR) is used in cognitive automobiles for interpreting and classifying traffic situations [Vacek et al., 2007]. The idea behind CBR is to solve a problem (case) by finding a similar previous situation and adapting it to the current situation. Afterwards, the new experience is integrated into the current database of cases for future usage. One big advantage of CBR methods is their support for generating partial matches between the current case and all stored cases. Vacek et al. use CBR to predict the evolution of the current situation in order to select appropriate behaviors for the automobile. Predicting future situations requires classification of situations or situation types, behaviors of participants, and risk potentials before drawing conclusions.

Hidden Markov Models (HMM) are another viable alternative for situation interpretation in cognitive automobiles [Meyer-Delius et al., 2009]. In this work, a situation is defined as a distribution over sequences of states associated with a meaningful interpretation. The distribution of each situation is described by an HMM which is also used to recognize the situation. The presented system tracks multiple possible situations in parallel, makes reasonable choices between competing hypothesis, and can be used to predict the position of all tracked vehicles.

HMMs can also be used to estimate the affective state of a human after being stimulated by robot motions [Kulic and Croft, 2007]. The approach is evaluated by letting a robot perform a pick up action and a reach and retract action close to a human. HMMs are used to estimate an arousal level or danger index based on physiological signals such as heart rate, skin conductance, and facial muscle contraction. This danger index is then used as input to the control system which adjusts the robot's movements to achieve a lower index and make the user more comfortable. In general, HMMs are used in situations where only the external influence of a system can be observed but their effect on the unobservable internal system state is to be represented. In this work, the unobservable internal state represents the affective state of the human.

Forecasting the load of a software architecture is presented in [Garousi et al., 2009]. The proposed method works on a software model which is defined in UML 2 and annotated with resource utilizations based on the UML-SPT profile presented in Section 2.3. Load forecasting requires a scheduled execution of the software and uses the specified SPT annotations to calculate resource utilizations at design time, based on the calculated schedule and simulated input values. This resource analysis technique allows detecting resource overuse and predicting resource utilization at design time but does not provide the means for runtime evaluation.

The Palladio Component Model (PCM) is designed to enable performance evaluation and prediction of software architectures at design time [Happe, 2008]. Colored Petri Nets (CPN) are used for predicting software performance based on a software model. The model is afterwards validated by comparing predictions of the model with simulations and measurements of application executions. The goal of this approach is to determine resource utilization and bottlenecks before the software is fully developed and deployed to production systems.

A methodology for answering performance queries at runtime based on PCM is presented in [Kounev et al., 2010]. The Hierarchical Queueing Petri Nets based performance model is generated dynamically at runtime and reflects the application configuration and workload. After its creation, the model can be queried for predicting the application's performance depending on expected application input.

2.6 Resource-Aware Systems and Algorithms

Gordon Moore formulated Moore's Law in 1965, based on his observations of the trends in chip manufacturing [Moore, 2006]. Specifically, this law says that the number of transistors available in a highly integrated circuit doubles approximately every two years. The International Technology Roadmap for Semiconductors (ITRS [4]) provides an outlook of future manufacturing technologies and shows numbers similar to the ones predicted by Moore's Law. Although the integration density of transistors has been slowed down since 2012, the roadmap still predicts the availability of hundreds to thousands of CPU cores on one chip within the near future (10-15 years).

Up to now there is no common understanding on how to efficiently use and program thousands of CPUs. Several research topics exist in this area which are presented next. The resource-aware approach of the Transregional Collaborative Research Center Invasive Computing (SFB/TR 89) is described in detail in Section 1.2, but belongs into this category, too. From an application programmers point of view, the main ideas are similar and often include system wide load monitoring and the expression of service level objectives

[4] `http://www.itrs2.net/`

within a program. However, major differences can be found regarding resource reservation policies of specific approaches.

2.6.1 Resource-Aware Operating Systems

The scalable many-core operating system ROS uses Space-Time Partitioning to provide guarantees to applications that resources are made available when they are requested [Klues et al., 2010]. ROS allows the programmer of an application and a system administrator to express the resource demands of an application in both space and time. For example, an application might want to use 4 processors and 50% of the available RAM for 75% of the time. Each application can also demand more resources than initially requested but the operating system provides only guarantees for the initial request. Resources in ROS are defined as anything which is sharable between applications such as CPU cores, RAM, caches, memory bandwidth, I/O device access, and many more.

OctoPOS is a resource-aware operating system which heavily relies on asynchronous, non-blocking system calls [Oechslein et al., 2011]. It is developed in the context of the Transregional Collaborative Research Center Invasive Computing (SFB/TR 89) and is designed to manage heterogeneous resources such as CPUs, CPUs with hardware accelerators, or specialized parallel hardware structured similar to GPUs. OctoPOS enables an application to express a performance metric. These metrics express the performance of the application in dependence on different combinations of resource quantities. Afterwards, an application can demand a specific set of resources at any point during runtime. If the requested resources are granted is negotiated upon at runtime, by using the performance metric to balance system wide resource utilization. Depending on the outcome of the negotiation, another set of resources is provided to the application which does not have to explicitly meet the original demand. The application can then either adapt its algorithmic parameters to the newly available resources, continue working with the existing resources, or it can try to renegotiate for more resources.

The SElf-awarE Computing model (SEEC) provides a runtime system designed to reduce programming effort in modern multi-core systems [Hoffmann et al., 2011]. SEEC is implemented as a set of Linux kernel libraries and allows application programmers to specify both goals and progress of their application. System programmers are provided with means to specify actions which can be performed by SEEC and the hardware to influence applications. During execution, each application is monitored in order to create application and system models. The SEEC runtime system attempts to optimize resource allocations based on an adaptive control system and the generated models. Thus, SEEC optimizes system behavior instead of requiring programmers to optimize their application for a specific system.

METE (Meeting End-to-End QoS) allows specifying Service-Level Agreements (SLA) between applications and the operating system to ensure dynamic allocation of resources [Sharifi et al., 2011]. The main objective of METE is to dynamically partition shared resources in multi-core machines to allow concurrent applications achieve their specified performance. Resources managed by METE are CPU cores, shared caches and off-chip bandwidth. A feedback-based control loop captures application behaviors at runtime and adapts resource demands by querying a centralized resource broker.

Tesselation is an operating system which introduces Adaptive Resource-Centric Computing (ARCC) [Colmenares et al., 2013]. Within Tesselation, resources are distributed among Quality of Service (QoS) domains called cells. Interfaces for resource adaptation and composition of QoS domains are realized on top of the cell abstraction. Resources are assigned to cells and resource allocation is adapted at runtime while scheduling of resources within a cell is performed via user-level scheduling. Thus, resource allocation is separated from actual resource usage. A separate Resource Allocation

Broker is responsible for distributing available resources to cells while simultaneously trying to satisfy global goals such as energy efficiency, met deadlines, or throughput of applications.

The Autonomic Operating System (AcOS) is an extension and specialization of the Linux and FreeBSD kernel schedulers [Bartolini et al., 2013]. Its goal is to perform automatic and adaptive resource allocation based on user-defined performance goals which are specified via Service-Level Objectives (SLOs). AcOS focuses on allocation of CPU and CPU bandwidth and provides means for temperature based limiting of provided resources. A closed control loop realizes the adaptive resource allocation by observing the system state via monitors, deciding based on adaptation policies, and finally acting based on the outcome of the policies. The observation in the control loop is performed by Heart Rate Monitors (HRMs) which provide performance measurements. Furthermore, the operating system schedulers are changed to allow making runtime changes to the resources allocated to an SLO-bound application.

2.6.2 Application Autotuning

Application autotuning automatically optimizes parallel applications through empirical searches. Autotuning removes the requirement of manual specification of Quality of Service application parameters by programmers.

XJava extends the Java language with tasks and parallel statements which are further used to drive an autotuning process [Otto et al., 2010]. These new high-level language constructs allow expressing parallelism explicitly and allow the compiler to extract relevant tuning parameters such as thread count, load balancing strategies, or number of pipeline stages. At runtime, this knowledge is exploited by setting the determined parameters based on context information gathered at runtime. The autotuner then systematically changes the tuning parameters and measures the application performance to find an optimal parameter configuration for the specific machine it is executing on.

An operating system wide and application independent approach for autotuning a combination of multiple applications is presented in [Karcher and Pankratius, 2011]. The presented autotuner is integrated into the Linux kernel and works without involving the user in the tuning process. Special system calls are provided to the application programmer to add tuning parameters and to start and stop measuring the application performance. An application's repetitively executed hotspot must be annotated with these system calls to provide information on where to insert measurement points and what parameters to vary. At runtime, operating system data such as application workloads and global system state is measured and fed into the autotuner. The tuning process then employs a simplex-based optimization algorithm which aims at determining a cross-process optimal parameter combination. Since the set of determined tuning parameters is specific to a dedicated computer platform, applications are automatically re-tuned at runtime when they are executed on a new platform.

AtuneRT provides an approach for online autotuning of computations running on GPUs [Tillmann et al., 2014]. Optimizing memory access patterns, workload balancing, and control flow minimization is required to write high performance GPU applications. *AtuneRT* addresses these points by optimizing the parameters thread count, per-thread workload, and the degree of loop unrolling of GPU bound applications. The online property of *AtuneRT* means that the tuning can be changed at runtime in order to address changes in the workload or the size of input data.

2.6.3 Organic Computing

The Organic Computing (OC) concepts are designed to deal with increasingly complex networked systems by introducing so called self-x properties such as self-organizing, self-configuring, self-adapting, self-healing, or self-optimizing [Müller-Schloer et al., 2011]. One goal is to guarantee adaptive and trustworthy behavior of systems whose complexity, autonomy, or dynamics prevent foreseeing their behavior at design time. The self-x properties allow deferring decisions to the runtime of an OC system. These decisions are made by a Controller which receives input from an Observer which in turn supervises the executing OC system. Additionally, a human operator can provide input by setting system specific goals from outside the system.

The Organic Robot Control Architecture (ORCA) presented in chapter 4.5 in the OC book ([Müller-Schloer et al., 2011]) is realized with OC principles. ORCAs main idea is that so called Basic Control Units (BCUs) provide the functionality for the robotic system and that additional Organic Control Units (OCUs) supervise those BCUs. BCUs sent health signals which attach uncertainties to sensor values. These uncertainties vary depending on the current environment or faulty sensors. Based on these health signals, OCUs determine the state of monitored BCUs at runtime and perform self-organizing, self-optimizing, or self-healing actions based on the monitored state. Possible actions to deal with anomalies, faults or unforeseen situations are activation or deactivation of BCUs, changing of BCU parameters, or changing BCU interaction patterns. Incremental online learning is used to optimize adapting behaviors based on monitored performance of the system

An implementation of ORCA for a six-legged walking robot is presented in chapter 5.7 of [Müller-Schloer et al., 2011]. Self-organization, self-reconfiguration and self-healing are the main principles employed in building the control architecture. Required BCUs range from sensing and single robot leg control, to reflexes for handling uneven terrain, up to a health-signal based path planner. Upon deviation of health signals from their regular range, an on-demand adjustment of the robot's behavior is performed by adjusting BCU parameters or by activating reflexes. In case of serious damage to a robot leg, self-reconfiguration through detaching the faulty leg is possible. It is shown that a considerable number of faults can be corrected without interrupting the robot albeit resulting in slower movement when legs have to be detached. In addition, a health status based planning algorithm is presented capable of adapting the plan if the robot's health changes. Thus, an initial plan of walking over bigger obstacles will be modified if the robot has lost a leg and can not climb big obstacles anymore.

2.6.4 Resource-Awareness in Robotics

So far, research towards resource-awareness in robotics has been done, but is not one of the major or hot topic as of now. Results from research towards resource-awareness are shown in this section.

An architecture for autonomous rovers is shown in [Castano et al., 2006]. Every time a new command (such as taking a measurement) is sent to the robot, the control software checks if enough resources such as memory or power are available to fulfill the command. Checking is performed after an action finishes, in order to take the currently available resources into account.

Generating resource constraint plans for a mobile robot was published in [Steck and Schlegel, 2010]. Through annotation of the robot software with non-functional properties it is possible to analyze resource usage and to generate static real-time schedules of all involved components.

A task-based approach for generating resource-constrained architectures for service robots is presented in [Park et al., 2012]. Starting with an initial action to perform, an architecture including all required components to perform the action is generated. Afterwards, the architecture is optimized by minimizing the usage of duplicate components and taking resource constraints into account. During this step, components with high resource usage are replaced by other components providing the same semantics but using less resources. Before the architecture is instantiated and executed, a simulation is run to ensure that all constraints are met.

Real-time reconstruction of contacts on a robotic skin is shown in [Muscari et al., 2013]. The algorithm depends on the Skinware middleware for reading and addressing the robotic skin sensors [Youssefi et al., 2015]. The reconstruction algorithm takes input from multiple force sensitive sensors and combines them into information about contact areas and direction from which forces are exerted. Real-time performance of the algorithm can be tuned to match the availability of resources by adjusting the sampling size at runtime. This concept is supported by the underlying middleware which supports adapting both the number and precision of currently active sensors.

2.7 Sampling-Based Motion Planning

Motion planning is used for calculating a collision free movement of an object or a robot from a starting position to a goal position. A wide variety of algorithms for solving this problem exist. However, in the humanoid robotics domain mainly two randomized approaches are used to solve complex planning problems. These algorithms are called Probabilistic Roadmaps (PRM) [Kavraki et al., 1996] and Rapidly-exploring Random Trees (RRT) [Kuffner and LaValle, 2000]. This work focuses on variants of the RRT approach, since it allows for efficient solving of single query planning problems while maintaining probabilistic completeness. RRTs are designed for single query path planning which is defined as finding a solution to single path planning problem as fast as possible and without preprocessing. Furthermore, RRTs are probabilistic complete, meaning they find an existing solution with any probability given sufficient time to execute.

The space in which an object can be moved is the configuration space (C_{space}) in motion planning. C_{space} is then further divided into the free space C_{free} and the space occupied by obstacles $C_{obstacle}$ ($C_{space} = C_{free} \cup C_{obstacle}$). Thus, the goal of motion planning algorithms is to find a path in C_{free} which is collision free by definition. Dimensionality of C_{space} and thus C_{free} depends on the problem formulation. C_{space} will have 6 dimensions (rotation and translation along each axis), if the problem is specified in cartesian space. If the problem is specified as movements of each joint of a robot, C_{space} will have as many dimensions as there are joints or degrees of freedom (DoF) in the robot. Computing C_{free} is very time consuming for robots with many DoF and is often not feasible.

Instead, the RRT algorithm builds a tree T of reachable configurations in C_{free} starting at the initial configuration ρ_{start}. For n iterations, a random pose ρ_{rnd} is sampled in C_{space} and a configuration ρ_n is created by extending the tree towards ρ_{rnd} by an amount ε. Each ρ_n is checked for collisions, to decide whether ρ_n belongs to C_{free} and should be added to the tree or if it belongs to $C_{obstacle}$ and should be discarded. T is returned after the algorithm finishes. This procedure is called `BUILD_RRT` and is shown in Algorithm 1.

Algorithm 1 The BUILD_RRT algorithm starts with an initial configuration ρ_{start}, a number of iterations n, and an extension amount ε. Every iteration, a random configuration ρ_{rnd} is created, the tree T extended towards ρ_{rnd} by ρ_n, and finally ρ_n added to the tree if it is valid (collision free). T is returned after the algorithm finishes.

```
procedure BUILD_RRT(ρ_start, n, ε)
    T.INIT(ρ_start)
    for k ← 1..n do
        ρ_rnd ← RANDOM_CONFIGURATION()
        ρ_n ← NEAREST_VERTEX(ρ_rnd, T)
        ρ_new ← EXTEND(ρ_n, ρ_rnd, ε)
        if VALID(ρ_new) then
            T.ADD_VERTEX(ρ_new)
            T.ADD_EDGE(ρ_n, ρ_new)
        end if
    end for
    return T
end procedure
```

One of RRTs properties is the rapid exploration of unvisited areas of C_{space}. This property is based on the fact that the area of the Voronoi region (Appendix B) of a vertex is proportional to the probability that the vertex is selected for extension [Kuffner and LaValle, 2000]. Vertices with more free space around them are therefore more likely to be extended, since their Voronoi region is comparatively bigger than the Voronoi regions of other nodes in the search tree.

Many improvements to the original RRT algorithm have been published since its first publication. All modifications can be separated into three functional groups according to their goals and are explained in detail in the following sections. First of all, it is possible to further improve the quality of a solution, once it was found. Furthermore, it is possible to make changes to the sequential RRT algorithm in order speed up finding solutions. Last, computation of the RRT algorithm can be sped up by parallelizing it to make use of multi-core CPUs or graphics processing units (GPU).

2.7.1 Improving the Quality of a found Solution

The algorithms presented in this section are based on the original *RRT* and are modified to refine and improve the quality of an existing solution.

RRT* provides asymptotic optimality in addition to probabilistic completeness [Karaman and Frazzoli, 2011]. Asymptotic optimality is described in the paper as "almost-sure convergence to optimal paths" and is also mathematically defined. This property of RRT* is achieved through two major modifications to the original RRT. First, the process of adding new nodes to the tree of RRT* is changed. Instead of connecting a sampled configuration to its nearest neighbor *nn*, an intermediate configuration *c* is created near the sampled configuration. Then, *c*'s vicinity is searched for a parent node containing the lowest cost (shortest path length) between start node and *c*. The edge between parent and *c* is added to the tree, if the respective line segment is collision free. The second modification optimizes overall path costs in the tree through a rewiring step which is executed whenever a new node *c* is added to the search tree. Every node *n* in *c*'s vicinity is checked, whether its path costs are lower if the path is redirected through *c*. If the path including *c* is cheaper and the new line segment between *n* and *c* is collision free, this line segment is added to the tree and the old connection to n is removed from the tree. This rewiring of the search tree ensures that nodes are reachable through a path with minimal costs.

Convergence towards the optimal solution of the planning problem is provided by the Informed RRT* algorithm [Gammell et al., 2014]. Every time a better solution s is found by the algorithm, the sampling domain C_{space} is reduced to $C_{space} \bigcap P_s$. P_s is defined as the ellipse with focal points ρ_{start} and ρ_{goal} and a polar diameter equal to the cost of s (see Figure 2.7). Thus, P_s contains all solutions within C_{space} with a cost less than or equal to the cost of s. Due to this modification, Informed RRT* always improves the path $\rho_{start} \rightarrow \rho_{goal}$ and does not spend time on improving paths between ρ_{start} and all other configurations $\rho \in C_{space}$.

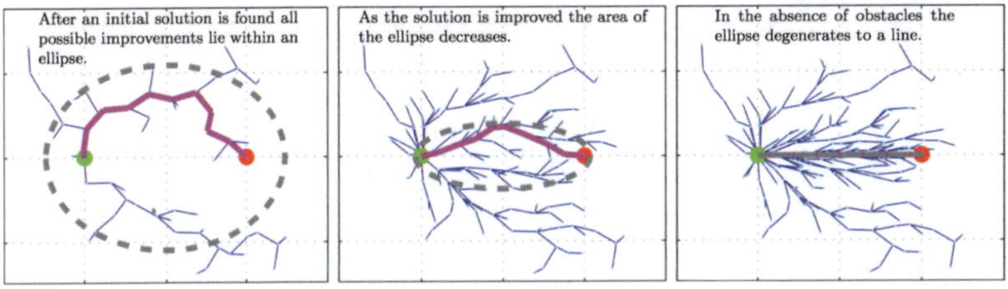

Figure 2.7: *Informed RRT** reduces the search space to an ellipse (gray dotted line) with focal points start (green) and goal (red). Each improved solution reduces the ellipse radius further. *Source:* [Gammell et al., 2014] ©2014 IEEE

RRT* Fixed Nodes (RRT*FN) enhances the RRT* formulation to restrict the memory available for storing nodes of the search tree [Adiyatov and Varol, 2013]. The motivation for limiting the memory for storing the search tree comes from the embedded systems domain where huge amounts of memory are not always available. To conserve memory, a node removal procedure is added to the RRT* formulation which removes weak nodes whenever new nodes are added to the tree. Nodes are considered weak, when their connection is cut during the rewiring of the tree or if they do not have any child nodes. The paper shows that RRT*FN comes close to computing paths which are almost as optimal as RRT* while needing much less memory.

Batch Informed Trees (BIT*) unifies techniques from graph- and sampling-based planning [Gammell et al., 2015]. The algorithm works in batches where each batch is responsible for finding a collision-free path from ρ_{start} to ρ_{goal}. Once a solution was found while executing a batch, a new batch is started in a search space restricted to an ellipsoid around the found solution. The planning search space itself is based on a graph of uniformly created samples which is used as a basis for a heuristic search. These properties allow the algorithm to converge towards the optimal solution similar to Informed RRT*.

2.7.2 Changing the RRT Algorithm to find Solutions faster

The original RRT formulation can be enhanced to guide the search towards the goal node. Different approaches are presented in this section.

RRTLocTrees provides a modification of RRT which is able to deal with narrow passages in the workspace of the robot [Strandberg, 2004]. If a valid sample can not be connected to the main search tree, a local RRT can be started at this sample. A heuristic determines if a new local search trees should be created. These local search trees are advanced together with the main search tree and a connection attempt is made, once a local search tree expanded its bounding box. Overall, the algorithm enables the exploration of narrow passages and finds paths through narrow passages faster than a regular RRT.

Blind RRT describes an enhancement of RRT which ignores obstacles in the workspace of the robot during an initial phase [Rodriguez et al., 2013]. This allows Blind RRT to efficiently explore the entire C_{space} at the expense of generating many disconnected trees. Those disconnected trees are merged again, after the initial exploration phase. The major advantage of this approach is a rapid expansion even if blocking obstacles are present in the scene.

Dynamic-Domain RRT adapts the search tree's sampling domain of nodes near obstacles by limiting their domain to an approximation of the currently visible Voronoi region called dynamic domain [Yershova et al., 2005]. Calculating the visible Voronoi region of a node in the search tree (see Figure 2.8(c)) is computationally intensive and should thus be avoided in motion planning. The dynamic domain of a node ρ is represented by a sphere around ρ with radius r_{border} (see Figure 2.8(d)) and is constructed as follows. Initially, each node in the search tree searches its total Voronoi region (Figure 2.8(b)), which is expressed by assigning each node a radius of ∞. Once connecting a node ρ_{rnd} to the nearest node ρ_n fails, the search space of ρ_n is reduced to its dynamic domain by setting the radius of ρ_n to the predefined value r_{border}. Nodes with a radius smaller than r_{border} are called boundary nodes and only configurations lying within their dynamic domain are chosen as valid samples if a border node is selected as a nearest neighbor. Thus, only border nodes have a smaller visible Voronoi region since they lie near the border of C_{free} and $C_{obstacle}$. In the presence of narrow passages in the workspace this property of boundary nodes reduces the number of collision checks and forces the *Dynamic-Domain RRT* algorithm to sample only in the boundary node's vicinity. Expansions towards far away and unreachable nodes is thus prevented. However, it is very important to choose an appropriate value for parameter r_{border} in order to properly approximate the visible Voronoi regions. It must be large enough to include enough of the node's visible Voronoi region but also small enough to exclude regions outside the node's visibility.

Adaptive Dynamic-Domain RRT enhances the original Dynamic-Domain RRT algorithm to make it more robust against poor choices of parameter r_{border} [Jaillet et al., 2005]. The original Dynamic-Domain RRT paper already explains that the introduced parameter r_{border} needs to be carefully selected to achieve good performance. The Adaptive Dynamic-Domain RRT addresses this problem by iteratively adapting the sampling domain of a node at runtime. After the radius of a node was set to r_{border}, this radius is decreased by a value of α for each additional failed connection attempt. However, the radius is increased by α again for each successful connection attempt. A minimum value of r_{min} for the radius is provided to preserver the probabilistic completeness of the algorithm. This automatic adaptation of the dynamic domain significantly increases the algorithm's robustness compared to the original Dynamic-Domain RRT, since the approximation of the visible Voronoi region is more accurate. This follows from the fact that border nodes with many obstacles in their vicinity have a lower radius than nodes surrounded by less obstacles.

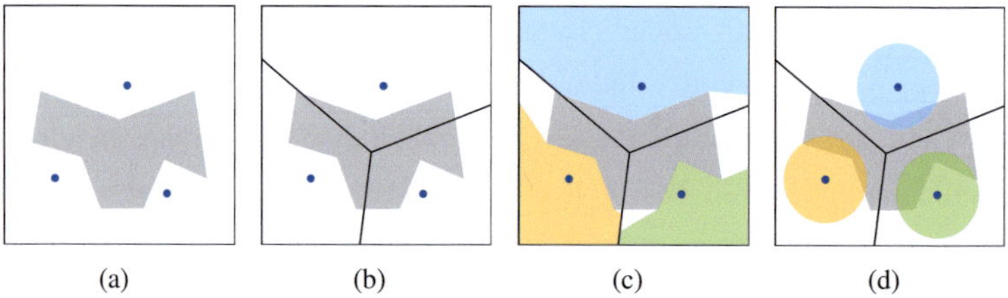

Figure 2.8: From left to right: a) three sampled points (blue) and an obstacle (gray), b) total Voronoi regions of the sampled points, c) visible Voronoi region of the sampled points, d) dynamic domains of the sampled points (spheres with radius r_{border}).

Adaptive RRT enhances the original RRT formulation with a selection process for expansion strategies. These strategies allow adapting the expansion of the RRT based on the available space [Denny et al., 2013]. The basic idea behind Adaptive RRT is that expansion of the tree is influenced by the available exploration area. An expensive expansion method might be beneficial for exploring narrow passages while inexpensive expansion methods are sufficient for open areas. Thus, the RRT formulation is enhanced with a mechanism for selecting an appropriate expansion strategy applied before sampling new configurations. Expansion strategy selection is based on the visibility of a search tree node which is constantly updated.

Poisson-RRT uses the maximal Poisson-disk sampling scheme to ensure good coverage of C_{free} and to generate the best distribution of samples in C_{space} [Park et al., 2014]. Using Poisson-disk sampling provides a guarantee that each new sample is a minimum distance r away from any existing search tree node. This is achieved by placing a hypersphere (or Poisson-disk) with radius r around each node in combination with the condition that no further samples can be placed inside the disk. Maximal Poisson-disk sampling uses these Poisson-disks and is defined as being finished, once no further Poisson-disk sample can be placed inside C_{space}. The properties of this sampling scheme ensure good coverage of C_{free} and the best distribution of Poisson-disk samples in C_{space}. In addition, the sampling scheme is enhanced with an adaptation mechanism which increases the sampling rate in narrow (difficult) regions of C_{space}. Once collisions between an edge of the sampling tree and obstacles occurred, the specific region is re-sampled using Poisson-disks with a smaller radius than r.

2.7.3 Parallelizing the RRT Algorithm

Reducing the planning time or increasing the planning problem difficulty requires additional computational power which can be provided through parallelization of the RRT implementation. The most relevant of the published approaches for parallelizing the RRT algorithm are presented next.

The Sampling-Based Roadmap of Trees (SRT) approach creates a global Probabilistic RoadMap (PRM) which uses multiple RRT instances to find solutions in locally confined spaces [Plaku et al., 2005]. As a result, the overall planner is more decoupled than single PRM or sampling-based tree planners which in turn allows distributing the computational workload evenly among processors and keeping communication costs low. SRT is also designed to answer multiple planning queries which goes beyond the scope of single query RRT's.

A basic OR parallel RRT is presented in [Devaurs et al., 2011, 2013]. This algorithm starts multiple independent RRT instances in parallel and finishes after one of the instances found a solution.

Manager Worker RRT collaboratively builds a common search tree in parallel [Devaurs et al., 2011, 2013]. To achieve this, the overall planning task is decomposed into two functionally independent subtask classes. The first kind of subtasks are operations requiring access to the search tree such as nearest neighbor search. The second type of subtasks do not require access to the search tree such as collision checking. All subtasks requiring access to the search tree are executed by a Manager process while all other subtasks are distributed to and collected from multiple Worker processes. This strategy works well as long as subtasks are independent from each other. However, the Manger Worker RRT becomes very inefficient if subtasks are intertwined as it is the case in the RRT* algorithm.

In contrast, the Distributed RRT makes use of exploratory decomposition in order to collaboratively build the search tree [Devaurs et al., 2011, 2013]. This approach distributes the planning onto multiple identical workers, each owning a copy of the sampling tree and performing sampling and collision checking. After every successful iteration, a worker sends its changes to the search tree asynchronously to all other workers. These changes are applied to the local search tree by all workers before starting their own next iteration. A stop signal is sent via broadcast once a solution is found and all workers terminate their calculations.

Both the pSBMP-RRT and pSBMP-PRM algorithm provide a parallel planning framework compatible with sampling-based algorithms such as PRMs and RRTs [Jacobs et al., 2012]. The major difference of these algorithms is the application of C_{space} subdivision to achieve superior scalability. Each C_{space} subdivision is processed in parallel and afterwards connected with nearby subdivisions. This results in a final solution covering the entire C_{space}.

Bulk Synchronous Distributed RRT is another parallel RRT implementation which modifies the Distributed RRT algorithm [Jacobs et al., 2013]. This algorithm addresses the communication overhead of Distributed RRT by broadcasting updates to the search tree only after a worker has finished computing a bulk of iterations. How the parallel expansion of the search tree works is shown in Figure 2.9. The same paper presents the Radial Subdivision Distributed RRT which subdivides C_{space} into conical regions. A search tree local to each region is constructed in parallel. Afterwards, neighboring regions are connected and existing cycles between regional trees are eliminated.

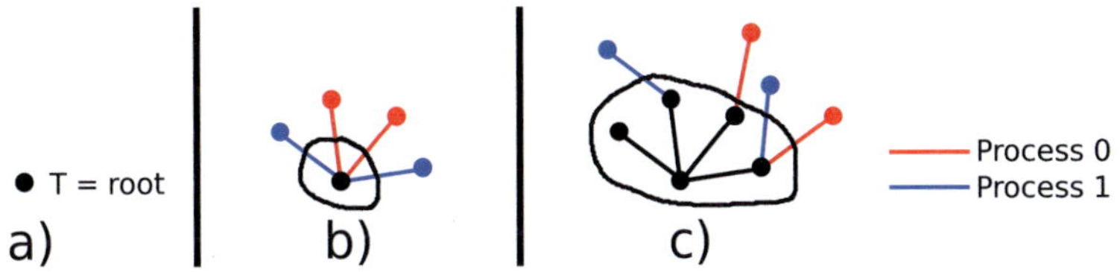

Figure 2.9: Bulk Synchronous Distributed RRT starts with multiple workers which only communicate after a certain amount of iterations. Two processes are shown (blue and red), each advancing the tree individually. After two iterations, the found solutions are synchronized and both processes continue planning with the new global view of the tree (Subfigure c)) [Jacobs et al., 2013].

The computationally most intensive part of RRT based approaches is collision checking, which is addressed in [Bialkowski et al., 2011]. Graphics coprocessors (GPUs) are used to perform a parallelized version of collision checking which in turn speeds up the basic RRT and the *RRT** algorithm.

Another possibility is to take advantage of cache effects and lock-free data structures as shown in [Ichnowski and Alterovitz, 2014]. The paper shows that lock-free data structures and cache friendly C_{space} subdivision can result in superlinear speedup for both RRT and *RRT** implementations.

This approach is however limited to multi-core processors sharing a common data cache.

RRTLocTrees spawns local search trees in narrow passages (see Section 2.7.2) [Strandberg, 2004]. The paper mentions that these local search trees can be executed in parallel, but no explicit evaluation is provided.

Radial Blind RRT applies the Blind RRT approach to the Radial RRT [Rodriguez et al., 2013]. In each radial subdivision of C_{space}, a Blind RRT is created and expanded in parallel. Afterwards, local trees are connected and cycles removed. The advantage of this approach is an overall reduction of global communication due to Radial RRTs local computations of the C_{space} subdivision and the improved exploration of blocked regions due to Blind RRTs properties.

Poisson-RRT and its enhancements to the RRT algorithm are described in Section 2.7.2 [Park et al., 2014]. The paper also presents a parallel implementation of Poisson-RRT which can take advantage of multi-core CPUs and GPUs. It is shown that Poisson-disk sampling results in fewer redundant nodes as opposed to the other parallel RRT algorithm implementation of the comparison. The Poisson-disk properties prevents multiple threads from generating redundant nodes which are too close to each other.

2.8 Summary

Overall, it was shown, that hardware architectures of all advanced humanoid robots are based on distributed computers and low-level controllers which are occasionally accompanied by special hardware. Depending on the hardware architectures, both control- and software frameworks follow a distributed design supported by the respective robotic middlewares. Robot applications built on top of those frameworks are usually designed as some kind of finite state machine or statechart. Even the concepts of the Task Description Language (TDL) can be mapped to statechart semantics since they share many similar ideas.

Contrary to the presented robot architectures, many different possibilities for modeling resources exist, all being specific to certain application domains. It is therefore necessary to select an appropriate model based on the use case at hand. The most detailed resource models can be found in PCM and the MARTE profile. PCM provides the possibility to describe the context a component is operating in. However, this context is specified statically at design time and contains the software environment the component will be run in.

In summary, a wide range of profiling and monitoring methods exist, ranging from low-level profiling on hardware level up to high-level plan execution monitoring normally used in combination with planning of robot tasks. Since planning is not a topic in this thesis, the focus is on low-level profiling and monitoring approaches, each coming with its own set of drawbacks. Some of them can only deal with single applications while others are designed to work on applications running on computers with hundreds and thousands of cores. Some measure very precisely but slow down the monitored applications and are therefore unusable for online monitoring, while others are less precise but do not work on all operating systems.

Different kinds of prediction models and algorithms are available for all kinds of use cases. Selecting an appropriate model depends on various parameters such as the type of information (discrete or continuous), the completeness of available domain knowledge, or the type of prediction. Different types of predictions must be chosen, if results should only be classified or if results should provide information about how a certain goal can be achieved. Thus, choosing an appropriate prediction model requires careful analysis of the application domain and the results desired from the prediction.

Several approaches for resource-aware systems have been presented, ranging from autotuning which requires minimal changes to an application, to approaches like specialized resource-aware operating systems or complete new paradigms which require adapting applications to the paradigm. Some of the presented approaches like auto-tuning change application parameters after executing an application while other approaches like Invasive Computing or Organic Computing allow dynamic changes while the application is running. This differentiation can also be seen in the publications from Section 2.6.4, where some robotic algorithms produce static allocations to satisfy resource constraints but can not deal with changes of available resources at runtime. On the other hand, the approach from [Muscari et al., 2013] shows the capabilities to adapt itself to dynamic changes at runtime.

Details about enhancements to the original RRT algorithm were shown to fit into three main categories. Namely, increasing the quality of a found solution, speeding up the sequential algorithm, and parallelizing the algorithm. However, no approach exists which combines all three kinds of enhancements. Furthermore, none of the parallel implementations takes the difficulty of the planning problem and availability of resources into account to dynamically modify the number of planning processes.

3 Speculative Resource Management

The key concept of speculative resource management is to improve resource allocation and distribution based on context-sensitive predictions of the resource utilization of future robot tasks, while taking the current state of the robot and its environment into account. With enough free resources available (or predicted), applications can be executed ahead of time in order to start computations before their results are required. Correct predictions and pre-computations can thus lead to shorter execution times without resulting in resource over-utilization. Additionally, external influences of the dynamic environment a robot operates in, can prevent the robot from performing tasks in the original order while requiring additional handling of such uncommon or unexpected events. In this case, correct anticipation of upcoming robot tasks and the associated resource utilization can prevent ahead of time execution of applications which are unused in the error handling case. Finally, information about predicted robot tasks and resources can provide the basis for detecting upcoming resource bottlenecks or conflicts and can be used to enhance the process of assigning and distributing resources.

This thesis focuses on two aspects of speculative resource management, which are explained in detail in the following sections. The first aspect is the data-driven generation of context-sensitive resource models for robot tasks based on data gathered from previous experience of former executions (Section 3.1). The second aspect is the context-sensitive prediction of future robot tasks and the associated resource utilization based on resource models (Section 3.2).

3.1 Data-Driven Generation of Context-Sensitive Resource Models

This thesis introduces a context-aware resource model which builds upon the definition of a basic resource model, which in turn depends on the applications to be described by these models. The application model influences the choice of required input and output parameters of the model as well as the level of abstraction on which to operate. Since this work is built on top of the ArmarX robot framework (Appendix C), the resource models are built on top of the framework's application model.

Context-sensitive resource models are required, since humanoid robots perform tasks in dynamically changing environments where robot tasks and associated resource usages differ based on the robot's surroundings. The context-sensitive resource model is created by associating the basic resource model with a context it is used in. For humanoid robots, this context consists of the environment the robot operates in as well as the actions the robot performs. It is also essential to find a suitable level of abstraction for influential environmental parameters in order to reduce the complexity of the model.

Model parameters required for building the model have to be extracted from executions of robot tasks by means of profiling and monitoring. However, acquisition methods differ widely, based on the type of data to extract. Thus, profiling and monitoring mechanisms must be constructed in a generic fashion in order to support various data types and their associated extraction technique. It is also essential for profiling and monitoring data of robot tasks to be accessible during execution and to be stored additionally for later processing. This allows building models at runtime or based on data from previous executions.

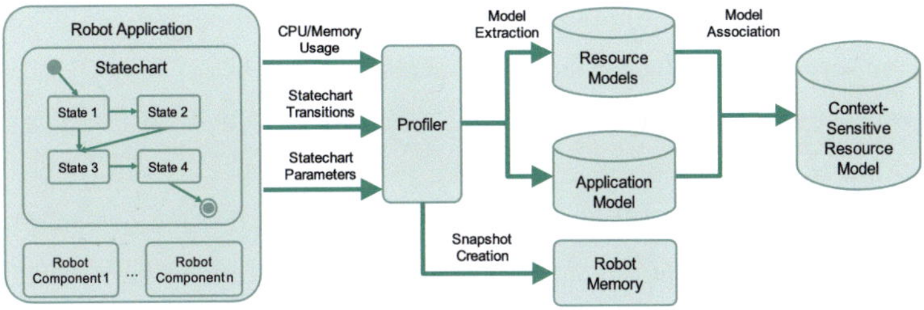

Figure 3.1: Overview of the profiling and resource model generation process. Robot tasks are composed of calculation performing robot components and control flow defining statecharts. During the execution of these tasks, robot component and statechart specific parameters are extracted by a *Profiler* and processed further into an application model and resource models. As a last step, both models are associated with one another to create context-sensitive resource models.

An overview of the overall profiling and resource model generation process is shown in Figure 3.1. The whole process is implemented using the ArmarX robot framework which is used to control the robots of the ARMAR series. New components such as the *Profiler* were introduced and existing components were enhanced to support profiling of application data.

3.1.1 Application Model

An ArmarX robot application consists of parts from two distinct functional categories. The first category contains so called robot components responsible for accessing hardware and performing computationally intensive operations. The second category comprises hierarchical statecharts containing high-level logic for coordinating robot components. Statecharts have been shown to be very descriptive and widely used for describing control flow and are therefore used both in ArmarX and as the main application model in this work (see Section 2.2). Additionally, ArmarX allows robot components and statecharts to be distributed over several network-connected PCs.

Robot components provide specific services within the ArmarX framework, such as retrieving camera images and sensor information, commanding motors, calculating robot poses, or detecting objects in camera images.

Different interface descriptions define input and output parameters of the specific services provided by the components.

ArmarX statecharts follow the design of the original Harel statecharts and allow hierarchical composition of states, meaning each state can be used as sub-state in another statechart. This enables the application designer to specify control flow at different levels of abstraction and allows building and reusing generic statecharts for different robot tasks.

Altogether, a statechart can be described as a tuple $(I, L, O, \Sigma, S, s_0, F, \delta)$. Each sub-state is defined as the same tuple with different parameters, while S, s_0, F, and δ are empty or undefined for leaf states.

A state without sub-states is called a leaf state and is defined by a set of input parameters I, a set of state local parameters L, a set of output parameters O, and a set of outgoing events Σ. Input parameters are used for state parameterization and are accessible during state execution. Output parameters can be thought of as return values of a state and are set during execution to values derived from internal calculations or data retrieved from robot components. Events are triggered based on conditions specified by the statechart developer.

A state containing a set of sub-states S additionally contains a start state $s_0 \in S$, a set of final states $F \subset S$, and a state transition function $\delta : S \times \Sigma \rightarrow S$. δ describes the transitions from the current state s_c to any state $s \in S$ based on the output events emitted by s_c. Each transition also maps parameters from different sources such as input parameters of the parent state or output parameters of the preceding state to the input parameters of the target state.

3.1.2 Basic Resource Model

Multiple possible definitions of resource models are presented in Section 2.3, ranging from abstract to very concrete models as well as from simple to more complex models. Robot tasks consist of concrete applications (robot components and statecharts) which are executed on physical hardware.

Therefore, the resource model of the "UML2 profile for Modeling and Analysis of Real-Time Embedded Systems" (MARTE) fits best since it provides a general approach to resource modeling [Object Management Group (OMG), 2011]. The Palladio Component Model (PCM) is similar in its resource description but the advanced features for describing component-based software architectures are not required in this use case.

The MARTE model is complex due to its generalized approach. In general, MARTE models are used to specify resource constraint parameters and non-functional properties (NFPs) at design time based on abstract resource types without connections to a concrete hardware/platform model. Once the design is finalized, a concrete instance of the hardware/platform model is selected which satisfies the specified constraints and NFPs.

The different resource types or entities defined in the MARTE profile are:

- *ExecutionHost*: a hardware device for executing operations.
- *CommunicationHost*: a hardware link between devices for exchanging data.
- *SchedulableResource*: a software resource managed by the operating system such as processes or threads.
- *CommunicationChannel*: a software-based middleware or protocol layer for transporting messages.

Each entity is associated with different resource related parameters which are described as a combination of a value and an associated SI-unit such as hours [h] or frequency [Hz]. The parameter value itself can either be expressed as a statistic measure (extracted from profiling data) or as different probability distributions. The resource related parameters defined in the MARTE profile are:

- *repetitions*: repetition count of a step or a loop.

- *probability*: probability of branching to a sub-path.
- *hostDemand*: host demand in time units (CPU requirement).
- *priority*: priority of a task, process or thread executed on a host system.
- *respTime*: time from starting a task until a result is returned (including initial scheduling delays).
- *executionTime*: time for executing a task or performing a calculation (excluding scheduling delays).
- *interOccTime*: time interval between two successive occurrences or executions.
- *throughput*: number of executions per time unit.
- *utilization*: the time an entity is busy divided by the mean number of busy copies ($100\% - utilization$ = the amount of time an entity is idle).
- *utilizationOnHost*: resource utilization of an entity on the host system (includes OS overhead).
- *memory*: amount of memory an entity uses during execution.

The use case in this work differs from the regular modeling with MARTE approach since the concrete hardware platform and the applications are given without any resource related parameters. These parameters must therefore be extracted from executions of the given applications on the existing hardware setup.

The platform model P describing the underlying computer architecture of the robot is a collection of PCs identified via CPU frequency, number of CPU cores, maximum memory, and network bandwidth

$$P := \{\ (\text{CPU\_frequency}, \text{CPU\_cores}, \text{maxMemory}, \text{networkBandwidth})\ \}.$$

The hardware/platform model for the humanoid robot ARMAR-IIIa consists of 4 PCs, excluding the gateway PC. All PCs are connected via Gigabit Ethernet and provide the following specifications:

- Coordination: 8GB RAM, Intel(R) Core(TM) i7-3770S @ 3.10 GHz
- Platform Control: 4GB RAM, Intel(R) Core(TM)2 Duo P9300 @ 2.26 GHz
- Hardware Control: 8GB RAM, Intel(R) Core(TM) i7-2715QE @ 2.10 GHz
- High-level Control: 4GB RAM, Intel(R) Core(TM) i7-3770S @ 3.10 GHz

From the list of resource parameters defined in the MARTE profile, *memory*, *respTime (repsonseTime)*, and *utilization* are the ones used for the basic resource model. These parameters are the most relevant ones for describing an existing system, as well as the memory consumption of a calculation, its duration, and if an application is compute-bound. Other parameters such as repetitions or priority are not included. They are more important for performing scheduling analysis at design time.

The resource model of a single robot component is thus defined as tuple

$$(\text{componentName}, \text{memoryUtilization}, \text{cpuUtilization}, \text{duration}, \text{platform}),$$
$$\text{platform} \in P.$$

The basic resource model is then defined as the collection of all resource model tuples of all robot components

$$RM_b := \{ (\text{componentName}, \text{memoryUtilization}, \text{cpuUtilization}, \text{duration}, \text{platform}) \mid \text{platform} \in P \}.$$

3.1.3 Context-Sensitive Resource Model

The developed context-sensitive resource model consists of basic resource models embedded into a robot specific context. For humanoid robots, this context consists of the robot's internal state (including the parameterized control flow) and the external state of the robot's environment. The internal robot state is represented on the ArmarX statechart level and is also required for recognizing the current robot task. Environmental parameters influence the sequence of robot tasks and thus the internal robot state. It is essential to describe these environment parameters on a high level of abstraction as well, since this helps to reduce the complexity of the resulting model. This is achieved through a symbolic representation of environmental objects which provides semantic information about the presence of objects stored in the robot's memory architecture. The PCM approach can not be applied in this case, since the PCM idea of a component's context is statically specified at design time and does not account for dynamic and unforeseen changes in the robot's environment.

Transitions within applications must be modeled in a probabilistic manner in order to describe the relationship between internal robot state and the environment. The MARTE parameter *probability* serves exactly the purpose of annotating transitions with probabilities. However, only one probability is assigned to each transition in the original MARTE description. This approach must be enhanced, since the decision for taking a specific transition is influenced by statechart parameters and by environment parameters.

For example, if a robot is ordered to fetch a bottle from the table, the most relevant input parameters for the fetching task are the object (bottle) and the pick up location (table). Based on the object type, an appropriate object localization algorithm can be chosen. Furthermore, the current robot location determines, if the robot is at the pick up location or if it has to move there first. Once the robot is at the pick up location, the robot task is influenced by the presence of the object to fetch and other objects acting as obstacles.

If the object is not on the table or unreachable, an alternative task has to be determined.

Altogether, the sum of observed contexts C can be described as a collection containing stateParameters and environment pairs

$$
\begin{aligned}
C := \{ & \, (\text{stateParameters}, \text{environment}) \mid \\
& \forall\, \text{sp} \in \text{stateParameters} : \text{sp} \in I \cup O \, \}.
\end{aligned}
$$

A robot statechart can then be defined as a collection of transitions T of the application model associated with a context and a probability. The probability can either be represented as occurrence frequencies or as probabilistic measures.

$$
\begin{aligned}
T := \{ & \, (\text{sourceState}, \text{destinationState}, \text{event}, \text{context}, \text{probability}) \mid \\
& \text{event} \in \Sigma, \text{sourceState} \in S, \text{destinationState} \in S, \text{context} \in C, \\
& (\text{sourceState} \times \text{event} \rightarrow \text{destinationState}) \in \delta \, \}
\end{aligned}
$$

This definition provides the means for reasoning about future state transitions. Based on the currently active state and context, exactly or partially matching transitions can be extracted from T.

The context-sensitive resource model RM_{cs} associates every transition $\in T$ with a collection of basic resource models $rm_b \in \text{RM}_b$ of robot components running during the execution of the destination state of transition

$$
\begin{aligned}
\text{RM}_{cs} := \{ & \, (\text{transition}, \{ \, \text{rm}_b \, \}) \mid \\
& \text{transition} \in T, \\
& \forall \text{rm}_b \in \text{RM}_b : \\
& \quad \text{rm}_b \text{ occurred during execution of transition.destinationState} \, \}.
\end{aligned}
$$

3.1.4 Profiling and Monitoring

Generating context-sensitive resource models from previous experience requires a profiling infrastructure which allows recording the internal robot state and the external environmental state. This profiling infrastructure must also be fast enough to support self-monitoring tasks in order to provide information about the current context for prediction and model update purposes. This section describes first, which parameters were selected to be profiled or monitored and which approach is used to extract the relevant information. Second, the profiling extensions made to the ArmarX framework are described.

As described in Section 3.1.2, it is essential to gather information about CPU and dynamic memory utilization of robot components as well as the execution time of robot tasks. Additionally, information about statechart executions and symbolic environment data is required to build the context for the context-aware resource model as described in Section 3.1.3. Statechart executions are described by transitions occurring between states, the time when a state is entered, input parameters as well as output parameters, and the time when a state is left. Information about the current environment is stored in the robot's working memory which is part of the memory architecture called MemoryX (Appendix C.2). MemoryX allows components of the robot application to perform queries on its complete contents, for example if certain objects are present in the scene. Furthermore, components can register with MemoryX to receive a notification for changes occurring in the memory, such as updates of an object position.

Several existing methods for extracting CPU utilization from applications were described in Section 2.4. Many methods such as Valgrind can not be used for extracting data, since they drastically reduce the execution speed of the profiled application and provide required information only after execution has finished. Thus, the only viable option is to retrieve the application CPU usage by querying the underlying operating system. Linux based operating systems provide the same CPU utilization data through both the `/proc` filesystem or the `getrusage()` programming interface.

Profilers such as Valgrind Massif provide reliable information about dynamic memory usage, but the same limitations as with measuring CPU utilization with Valgrind apply. Querying the operating system only provides reliable information about an application's static memory. However, dynamic memory information can only be retrieved by querying the memory allocator of a programming language. An operating system can not know about the currently occupied dynamic memory since it only hands out virtual memory to the allocator. The allocator then actually allocates dynamic memory chunks when requested by the application. On Linux based systems running the GNU C library, this dynamic memory information can be retrieved by calling the `mallinfo()` utility function.

One core principle of ArmarX is disclosure of the system state which allows querying ArmarX robot component parameters at runtime. It is therefore possible to poll input and output parameters of statecharts and the currently active state at runtime. Major drawbacks of polling are high bandwidth utilization if no parameters changed and missed updates if the polling frequency is too low. To address these issues, it is essential to enhance the ArmarX statecharts to report internal values upon change.

Introspecting the content of the robot memory is also possible due the system state disclosure principle of ArmarX. In order to reduce the amount of memory information to process, snapshots of the current environment should only by generated at crucial points in time such as occurrence of transitions in a statechart. To achieve this, it is essential to couple generating memory snapshots with reporting ArmarX statechart changes.

Profiling Architecture

An ArmarX application consists of one or more robot components responsible for performing tasks such as reading sensor values, setting control signals, or recognizing objects. A special statechart component exists for instantiating and executing statecharts. Each state is managed by a StateController which is responsible for triggering transitions between sub-states based on internal or external events and for managing input and output values of the sub-states. Each sub-state is again managed by a new instance of a StateController.

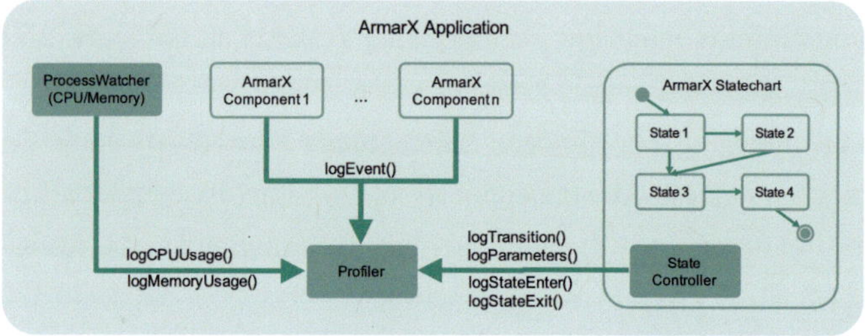

Figure 3.2: Overview of the ArmarX profiling architecture. Each application contains a single Profiler instance. The ProcessWatcher is responsible for measuring CPU and memory utilization and forwarding it to the Profiler. Each Component is able to log events via the Profiler and every statechart informs the profiler of occurring transitions (entering or leaving of states), and current state parameters.

The ArmarX framework was enhanced by several mechanisms to realize the profiling and monitoring architecture. See Figure 3.2 for an overview of the main parts of the profiling architecture. Each ArmarX application holds a single instance of the abstract Profiler class which contains methods for gathering data, collecting events, creating timestamps, and for retrieving the Process ID of the current application. Several specialized implementations of the Profiler exist for storing received data in files, for sending received data over the network, and a version for sending bulk data over the network. The Profiler is accessible from anywhere in an ArmarX application, can be activated and deactivated selectively, and allows straight-forward interface extensions for dealing with future requirements.

Data is received by the Profiler from different sources depending on the data type. Memory and CPU utilization is captured by a ProcessWatcher executing periodically every 300 ms in a separate thread in each ArmarX application. On Linux based operating systems, CPU utilization is retrieved by the ProcessWatcher through parsing the application specific CPU statistics from `/proc/${ProcessID}/stat` files. The left image of Figure 3.3 shows an exemplary CPU utilization profile for different robot components. Simultaneously, the C library memory allocator is queried for the current status of occupied dynamic memory. The right part of Figure 3.3 shows an example of a dynamic memory utilization profile.

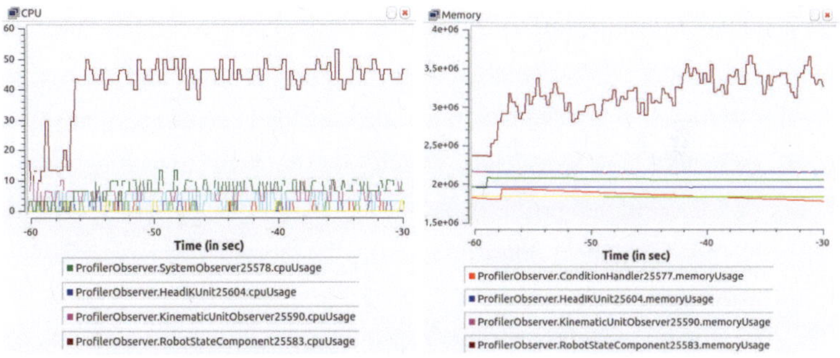

Figure 3.3: Plots of CPU and memory utilization profiling data. On the left, CPU utilization in % is shown and on the right, dynamic memory utilization in MByte.

Statechart profiling data is gathered in the scope of an enhanced State-Controller which calls the Profiler when state events occur. A transition is identified by a timestamp, source state, destination state, and the triggering event. On entering a state, its state identifier, a timestamp, and a FUNCTION_START event is recorded in combination with all state input parameters and finally all state local parameters. When a state is left, a FUNCTION_EXIT event is recorded together with a timestamp, the state identifier, all state local parameters, and all state output parameters. Additionally, it is possible to explicitly specify the hierarchy level at which profiling should be stopped. This feature is especially helpful for excluding sub-states in which high transition rates would result in huge amounts of profiling data.

The only parts of the profiling architecture not shown in Figure 3.2 are the dedicated robot component for creating memory snapshots and the ProfilerStorage component. Snapshot creation can not be integrated into every ArmarX application, since it depends on executing statecharts and requires access to the MemoryX components holding the external environment state. Therefore, the snapshot component receives state, transition, and parameter information from the Profiler instances of the statechart components and queries the working memory component, once a state transition is reported. Already existing snapshots are reused while non-existent snapshots are created and stored permanently in MemoryX. Storing new snapshots can occur multiple times per second depending on the profiled statechart. Hence, the snapshot component is designed in a multi-threaded and concurrent way to be able to handle high workloads. ProfilerStorage provides an interface to MemoryX for storing all profiling data which is received by the Profiler. Additionally, this component allows retrieving the stored information for later offline processing.

The presented profiling architecture can additionally be used for monitoring purposes, if a Profiler is enabled which sends out the profiling data via network. This enables the generation of new or updated resource models at runtime.

3.1.5 Resource Model Generation

The resource model generation process extracts models from profiling data as shown in Figure 3.1. It starts by generating data through profiling of robot tasks and creating environment model snapshots. From this data, a probabilistic application model is generated, based on which the intervals for the basic resource models are determined. The last step associates the basic resource models with the application model to create the context-aware resource model. The complete process is described in Algorithm 2 and the called high-level functions are specified in more detail in the later algorithms.

Algorithm 3 describes how durations of state executions are calculated by finding matching *enter* and *exit* events. Algorithm 5 describes how basic resource models are generated by splitting CPU and memory utilization data based on components and execution intervals and how component dependent statistics for CPU and memory utilization are calculated. Afterwards, these statistics are associated with the state which executed during the interval by assigning them to a state name in combination with the timestamp at which the computation started. Finally, Algorithm 4 describes how data of the previous two algorithms is used to create the context-sensitive resource model by finding similar transitions, comparing them with transitions executed in the same context, and by finding all resource models within the respective execution intervals.

Algorithm 2 The algorithm for offline generation of the context-sensitive resource model RM_{cs}. CALCULATE_STATE_TIMINGS is explained in more detail in Algorithm 3, GENERATE_BASIC_RESOURCE_MODELS in Algorithm 5, and GENERATE_CONTEXTSENSITIVE_RM in Algorithm 4.

```
procedure GENERATE_RESOURCE_MODEL()
    (EventList, CPUList, MemoryList) ←
        READ_DATA_FROM_MEMORY()
    SORT_BY_ASCENDING_TIMESTAMP(CPUList)
    SORT_BY_ASCENDING_TIMESTAMP(MemoryList)
    // split events in EventList by ProcessID
    for event ∈ EventList do
        ProcessID ← event.ProcessID
        EventMap[ProcessID] ← EventMap[ProcessID] ∪ event
    end for
    for element ∈ EventMap do
        // sort by timestamp:
        // element.key contains a ProcessID
        // element.value contains a list of ProcessID associated events
        SORT_BY_ASCENDING_TIMESTAMP(element)
    end for
    StateTimingMap ←
        CALCULATE_STATE_TIMINGS(EventMap)
    StateResourcesMap ←
        GENERATE_BASIC_RESOURCE_MODELS(
                        StateTimingMap, CPUList, MemoryList)
    RM_cs ←
        GENERATE_CONTEXTSENSITIVE_RM(
                                        StateResourcesMap)
    return RM_cs
end procedure
```

Algorithm 3 CALCULATE_STATE_TIMINGS receives a map containing (ProcessID, EventsForProcessID) pairs and calculates the execution duration of states. These durations are returned in map containing (stateName, timingInformation) pairs.

```
procedure CALCULATE_STATE_TIMINGS(EventMap)
    // calculate durations of state execution
    for eventList ∈ EventMap do
        for stateEnterEvent ∈ eventList do
            stateExitEvent ←
              FIND_CLOSEST_EXIT_EVENT(
                                              stateEnterEvent)
            duration ←
              stateExitEvent.timestamp − stateEnterEvent.timestamp
            timing ← (stateEnterEvent.timestamp,
              stateExitEvent.timestamp, duration)
            StateTimingMap[stateEnterEvent.stateName] ←
              StateTimingMap[stateEnterEvent.stateName] ∪ timing
            // remove processed events from EventList
            eventList ← eventList ∩ {stateEnterEvent, stateExitEvent}
        end for
    end for
    return StateTimingMap
end procedure
```

Algorithm 4 GENERATE_CONTEXTSENSITIVE_RM receives a mapping between state names and associated basic resource models. It builds the probabilistic application model and the context-sensitive resource model by attaching the resource models from the received mapping to the new application model.

```
procedure GENERATE_CONTEXTSENSITIVE_RM(
                                        StateResourcesMap)
    TransitionData ←
      READ_TRANSITION_DATA_FROM_MEMORY()
    // generate application model from transition statistics
    for transition ∈ TransitionData do
        similarTransitions ← {t|
            t.event = transition.event,
            t.sourceState = transition.sourceState,
            t.destinationState = transition.destinationState}
        sameContextTransitions ← {t|
            t ∈ similarTransitions,
            t.context = transition.context}
        // transition probability of states with equal contexts
        transition.probability ← |sameContextTransitions| / |similarTransitions|
        transitionTimes ← {t.timestamp|t ∈ sameContextTransitions}
        // remove processed transitions from TransitionData
        TransitionData ← TransitionData ∩ sameContextTransitions
        // find relevant basic resource models
        resources ← {r.value.resourceModel|
            r ∈ StateResourcesMap,
            r.value.startTime ∈ transitionTimes,
            r.key = transition.destinationState}
        // connect application model with the basic resource models
        RM_cs ← RM_cs ∪ (transition, resources)
    end for
    return RM_cs
end procedure
```

Algorithm 5 GENERATE_BASIC_RESOURCE_MODEL builds basic resource models based on state timing information (StateTimingMap) and data from CPU (CPUList) and memory profiling (MemoryList). The result consists of a mapping between stateName and (startTime, resourceModel) pair. CALCULATE_STATISTICS takes a list of data points and calculates the minimum, maximum, mean, and standard-deviation. The algorithm listing continues in Algorithm 6.

```
procedure GENERATE_BASIC_RESOURCE_MODEL(
                          StateTimingMap, CPUList, MemoryList)
    /* create basic resource model */
    for timingList ∈ StateTimingMap do
        for time ∈ timingList do
            // select CPU and memory utilization for execution interval
            cpuUsage ← {cpu|
              cpu ∈ CPUList,
              time.startTime ≤ cpu.timestamp ≤ time.endTime}
            memoryUsage ← {memory|
              memory ∈ MemoryList,
              time.startTime ≤ memory.timestamp ≤ time.endTime}
            // split CPU and memory utilization by component
            for cpu ∈ cpuUsage do
                componentName ← cpu.componentName
                cpuComponentMap[componentName] ←
                    cpuComponentMap[componentName] ∩ cpu
            end for
            for memory ∈ memoryUsage do
                componentName ← memory.componentName
                memoryComponentMap[componentName] ←
                    memoryComponentMap[componentName] ∩ memory
            end for
Algorithm description is continued in Algorithm 6.
```

Algorithm 6 Second part of GENERATE_BASIC_RESOURCE_MODEL (Algorithm 5).

```
            // calculate resource statistics for each interval
            for cpu ∈ cpuComponentMap do
                componentName ← cpu.key
                platform ← cpu.values[1].platform
                // Store as (utilization, platform) pair
                CPUMap[componentName] ←
                    (CALCULATE_STATISTICS(cpu.values),
                        platform)
            end for
            for memory ∈ memoryComponentMap do
                componentName ← memory.key
                platform ← memory.values[1].platform
                // store as (utilization, platform) pair
                MemoryMap[componentName] ←
                  (CALCULATE_STATISTICS(memory.values),
                        platform)
            end for
            for componentName ∈ componentNames do
                memoryUtilization ←
                    MemoryMap[componentName].utilization
                cpuUtilization ← CPUMap[componentName].utilization
                platform ← cpuUtilization.platform
                RM_b ← RM_b ∪
                    (componentName, memoryUtilization, cpuUtilization,
                        time.duration, platform)
            end for
            // associate (startTime, resourceModel) pair with stateName
            StateResourceMap[time.stateName] ←
              StateResourceMap[time.stateName]
              ∪ (time.startTime, RM_b)
        end for
    end for
    return StateResourceMap
end procedure
```

After describing the resource model generation, the remaining step is to select an appropriate abstraction level at which the profiling of the robot application should be performed. Depending on the abstraction level, the frequency with which data is generated differs widely.

Low-level control algorithms of humanoid robots usually operate at a frequency of $1/,kHz$ or more [Ott et al., 2006, Diftler et al., 2011, Nelson et al., 2012, Englsberger et al., 2014], On the other hand, algorithms on a higher level of abstraction operate at a lower frequency. For example, the vision algorithms on the humanoid robot ARMAR-III are bound by the update frequency of $30Hz$ of the camera system [Asfour et al., 2006]. The focus is therefore set on profiling high-level behavior and control flow, since the goal of this thesis is to operate on the level of robot tasks. Profiling low-level control algorithms is omitted due to the high update frequencies and the fine granularity. Currently, expert knowledge is required to specify the important high-level states of the statechart.

3.2 Context-Sensitive Resource Prediction

The second aspect of speculative resource management addressed in this thesis is the prediction of future robot states and their associated resource utilization. An overview of the proposed prediction architecture is shown in Figure 3.4. This chapter presents a prediction model for robot tasks based on Markov chains (see Appendix A for mathematical details). The model is capable of learning and adapting the model parameters from previous executions and is used to predict the behavior of the humanoid robot ARMAR-III while performing human robot interaction tasks. Context and robot state information required by the prediction algorithm are provided by the monitoring concept introduced in Section 3.1.4. The preview of potential future resource demands is expressed as a probability distribution, which is created by combining predicted robot tasks with the context-sensitive resource models as described in Section 3.1.3. The implementation of the prediction model and the initial version of the prediction architecture originate from the master thesis of Tobias Haaß [Haaß, 2014].

In the context of a resource-aware robot framework or an invasive runtime system, this resource prediction provides additional information which can be considered during resource negotiation and allocation to optimize system performance by avoiding upcoming resource bottlenecks. Furthermore, predicting future resource usages allows for speculative resource allocation for upcoming tasks or for starting tasks earlier, if the resource utilization permits so. In case of correct predictions, speculatively allocated resources can decrease execution times, since the delay for allocating resources can be omitted. Similarly, starting tasks earlier can also reduce execution times if the prediction was correct. If earlier started tasks generate unusable results, due to incorrect predictions, those results will be discarded and calculated anew. Execution time is not impaired by incorrect predictions, if only tasks are started early, whose execution characteristics match available resources and who support adjusting to changes in available resources. Restricting the number of early started tasks and terminating tasks, which were started based on incorrect predictions, must be handled by the same mechanism which performs the speculative resource allocation.

Predicting future resource usages consists of multiple steps, as shown in Figure 3.4. First, the execution of the robot application must be monitored in combination with the environmental context. Next, the monitoring information must be used by a prediction model in combination with the application model described in Section 3.1.1 in order to calculate a probability distribution for possible future robot tasks (states in the statechart). Finally, associating the state prediction output with the previously generated context-sensitive resource models leads to a distribution of most likely future resource utilization possibilities.

Monitoring is performed via the *Profiler* component as described in Section 3.1.4. Every time a statechart transition occurs at runtime the *Profiler* reports the current state $s_c \in S$ of the robot's application model.

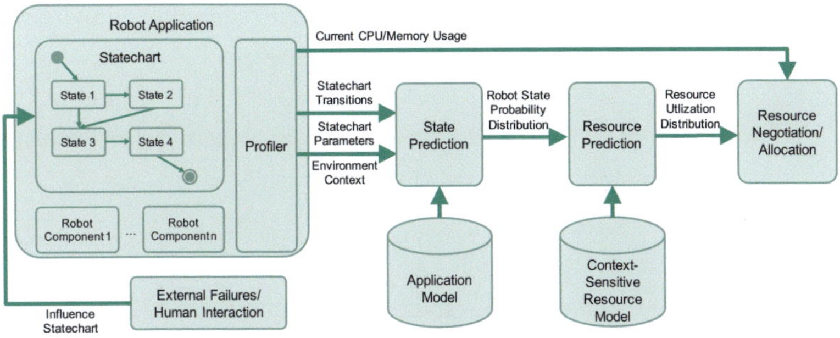

Figure 3.4: Overview of the resource prediction process. First, the *State Prediction* component calculates the prediction of future robot tasks. The result is consumed by the *Resource Prediction* component and can then be used to influence decisions on resource allocation or resource negotiation.

Additionally, transitions between all monitored states are reported as well as the current context $c_c \in C$. The context is defined as in Section 3 and consist of a snapshot of the environment the robot operates in, the tasks the robot performs, as well as associated input and output state parameters. Counting the occurrences of all transitions further leads to the transition probabilities expressed in the context-sensitive resource models.

3.2.1 State Prediction Model

Characterizing the environment is essential for the selection of a matching prediction model. In this work, the environment is described by the following properties:

- *dynamic*: changes in the environment are possible without any interaction of the robot.
- *discrete*: the states of environmental objects and the states of the robot statecharts are discrete.

- *stochastic*: external events such as a human entering and modifying the scene may occur with certain probabilities.
- *completely observable*: the state of all scene objects can be observed and is therefore known.

The goal of the state prediction model is to calculate future robot tasks, as in states of a statechart based on the current robot state and the current context or world state. A world state $\omega_i \in \Omega$ is similar to a transition $t \in T$ but contains only the context c_i and the current state s_i as parameters

$$\omega_i = (c_i, s_i),\ c_i \in C, s_i \in S.$$

Predicting the next world state ω_{t+1}, which is reachable from the current state ω_t, is defined as a function

$$p(\omega_t) = \omega_{t+1},\ \ \omega_t, \omega_{t+1} \in \Omega.$$

The set of possible world states $\omega_i \in \Omega$ is finite, since the environmental model for the prediction use-case in this work is defined as being discrete and completely observable. It is therefore possible to define a prediction model as a first order Markov chain.

The transition probabilities π_i from the current world state ω_i to every other world state $\omega_j \in \Omega$ is described by the probability vector $X \in \mathbb{R}^n$. The dimension n of X represents the number of observed world states, while X holds the following properties

$$\sum_{i=0}^{n} \pi_i = 1,\ \ \pi_i \geq 0.$$

The transition probability $p_{i,j}$ from world state ω_i to ω_j is defined as

$$p_{ij} := P(X_{t+1} = \omega_j | X_t = \omega_i),\ \omega_i, \omega_j \in \Omega; t \in T.$$

This results in the overall transition matrix M, which describes transition probabilities from any world state ω_i to any other world state ω_j. M is defined as

$$M = \begin{pmatrix} p_{11} & \cdots & p_{1n} \\ \vdots & \ddots & \vdots \\ p_{n1} & \cdots & p_{nn} \end{pmatrix}.$$

Each probability p_{ij} is independent of the time parameter t for any given X_t. This allows defining the prediction model as a homogeneous first order Markov chain. Furthermore, the homogeneous nature of the Markov chain allows calculating predictions not only for one future time step but for a prediction horizon of h future time steps. However, the greater the prediction horizon h, the less reliable the prediction outcome.

Calculating the prediction using a Markov chain given a world state ω_i requires the following steps

1. Initialize vector X_i with all entries set to 0 and mark the entry of ω_i (index i) as 1. X can thus be written as

$$X_i = \begin{bmatrix} \delta_{1i} \\ \vdots \\ \delta_{ni} \end{bmatrix}. \tag{3.1}$$

2. Construct the transition matrix M from the latest database entries according to

$$p_{ij} = \frac{count(\omega_i \to \omega_j)}{\sum_{x \in \Omega} count(\omega_i \to x)}, \tag{3.2}$$

where p_{ij} is the number of transitions from ω_i to ω_j divided by all transitions starting at ω_i.

3. Calculate the probability distribution for ω_j

$$X_j = X_i^T \cdot M.$$

4. The entry of X_i with the highest probability is chosen as the next most probable world state from which the related robot state is extracted.

5. Repeat calculating ω_j until the number of iterations matches the prediction horizon h. During each iteration replace the state vector X_i with the state vector X_j from the previous iteration.

6. Return the sequence of calculated X_i vectors.

Using only the above approach, it is impossible to calculate a prediction if an unrecognized world state is encountered, since no entry for this world state is present in the matrix M. Thus, a replacement strategy is employed which behaves like the Markov chain approach but without including the context parameters.

3.2.2 Resource Prediction

Resource utilization is predicted after the probability distribution of states and contexts are calculated. First, the most probable sequence of transitions $t_1, \ldots t_h$ is extracted from the prediction data. For each t_i the associated basic resource models are retrieved from the context-sensitive resource model RM_{cs}, see Section 3.1.3 for details. This list of resource models holds the information about the expected resource utilization in each state of the transition sequence $t_1, \ldots t_h$.

This procedure is repeated for the less likely transition sequences in order of descending probability, until their accumulated probabilities lie above a certain specified threshold. The result is a collection of the most probable state transitions which are annotated with the resource utilization of all components involved during the execution of these states. Furthermore, extracting minimum and maximum values of the resources specified in these models provides information about the expected range of resource utilization.

3.2.3 Online Learning and Updating

Initially, no input data for the prediction model is available and occurring transitions are regarded as uniformly distributed, regardless of the world state. All monitored statechart transitions are continuously stored in MemoryX in combination with a context snapshot. Upon first occurrence of a transition, missing information in the prediction model is generated based on the assumed uniform distribution, thus closing gaps in the model on the fly. Immediately taking these newly observed executions into account leads to continuous updates of the transition probabilities stored in the model matrix M.

3.2.4 Resource Prediction Architecture

This section describes the resource prediction architecture built on top of the profiling/monitoring architecture presented in Section 3.1.4. As indicated in Figure 3.4, the main components StatePrediction and ResourcePrediction are supplemented by the PredictionStorage component. As indicated by its name, PredictionStorage receives data of calculated predictions and stores it in MemoryX for later retrieval and evaluation.

Predicting state transitions and resource utilization in an ArmarX robot application requires starting the three components StatePrediction, ResourcePrediction, and PredictionStorage in combination with monitoring components described in Section 3.1.4. Every time a statechart transition is reported by the Profiler, the StatePrediction component is triggered. Based on the new current state and the current memory snapshot, a prediction is calculated and the probability distribution is broadcast to listening components such as PredictionStorage and ResourcePrediction.

The StatePrediction component is designed to support different prediction mechanisms. Figure 3.5 shows the connection between the StatePrediction component and the PredictionMethod interface it uses to calculate predictions.

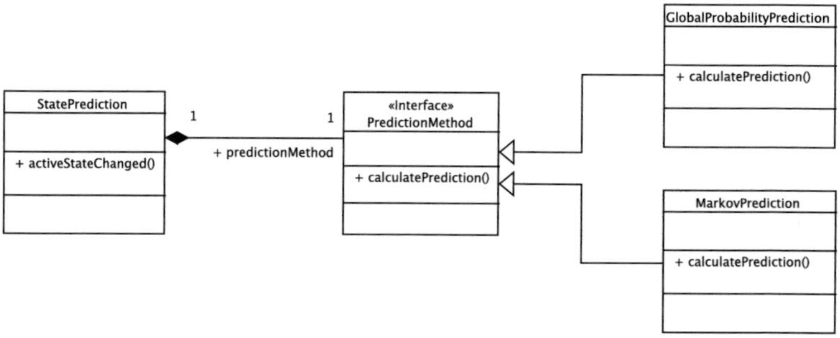

Figure 3.5: Relation between StatePrediction and the different prediction methods.

Once a new prediction method implements the PredictionMethod interface, it can be used by the StatePrediction interchangeably. As indicated in Figure 3.5, there are two implementations of the interface: MarkovPrediction and GlobalProbabilityPrediction. The MarkovPrediction performs the calculation as described in Section 3.2.1. This method uses the ProfilerStorage component to access the data required to construct the prediction model consisting of the transition probability matrix M. The size of M is $n \times n$ for n observed situations. In order to handle model data changes, the matrix is created and updated on the fly. Equation (3.2) is used for calculating the individual probabilities stored in the matrix. Before a prediction is computed, a world state ω_i matching the current world state is computed and the initial vector X_0 is created according to equation (3.1). Computing a matching world state ω_i is achieved by combining the current active state s_c with a current memory snapshot and comparing it with already stored world state instances. If no matching world state can be found, the fallback method is used for computing the prediction. The GlobalProbabilityMethod (GP) is another implementation of the PredictionMethod interface. This implementation relies solely on global transition probabilities and discards any kind of environment or statechart parameters.

After a new prediction is broadcast by the StatePrediction component, the ResourcePrediction component is triggered. ResourcePrediction uses the predicted probability distribution to generate a resource utilization distribution as described in Section 3.2.2. The context-sensitive resource model required for this process is retrieved from MemoryX and the prediction horizon parameter h is defined at component startup. Each resource prediction contains information about the most likely, the worst case, and the best case changes in resource utilization. The finished resource prediction is then broadcast to any listening component, such as the PredictionStorage, which creates a persistent copy of the data in MemoryX, or the speculative resource allocation framework.

4 Resource-Aware Algorithms

This chapter addresses the question of how algorithms can benefit from resource-aware concepts. Two different approaches are described for two different kinds of algorithms. First, a motion planning algorithm is presented which is capable of dynamically adapting its resource utilization based on the planning problem difficulty. The algorithm's performance is close to a parallel algorithm executed with a fixed amount of resources while providing better CPU utilization. Additionally, no specialized hardware is required since the implementation runs on standard PC hardware. The second resource-aware algorithm calculates a depth map from a pair of stereo images. This algorithm is based on mechanisms developed in the Transregional Collaborative Research Center Invasive Computing (SFB/TR 89) [Teich et al., 2011]. Calculations are distributed across multiple cores while size and amount of the distributed chunks are calculated based on the available resources.

4.1 Resource-Aware Motion Planning

Motion planning in cluttered environments is a difficult task, for which randomized motion planning algorithms are known to provide solutions in and adequately fast time. Parallelization further improves the problem solving speed of randomized algorithms such as the Rapidly-exploring Random Tree (RRT). Best performance for solving difficult planning problems is achieved when parallelized motion planning algorithms utilize all available computing resources.

However, humanoid robots must execute control, perception, and decision making algorithms in addition to motion planning. All these algorithms need to be scheduled and distributed concurrently onto the limited on-board resources of the robot (CPU, memory, communication bandwidth, power) while additionally being constrained by the limited available battery power. In this context, better resource utilization and allocation on system level is facilitated by resource-aware algorithms which are capable of adapting their resource demands to the dynamically changing system load.

This section presents a parallelized distributed motion planning approach which is able to request and release resources at runtime in a resource-aware manner. Figure 4.1 shows the overall architecture of this motion planner. Allocation strategies perform dynamic resource allocation and algorithm adaptation based on the current planning progress and an approximation of the planning problem difficulty. The approximation is calculated at runtime through different self-monitoring strategies. Compared to algorithms with static resource allocation, this approach can reduce system workload and still meet Quality of Service (QoS) measures such as average workload or efficiency. This resource-aware motion planner originates from the bachelor thesis of Raphael Grimm [Grimm, 2015].

The algorithm is designed to operate inside a resource-aware robot framework which knows about the system status and resource utilization of all running components. Resource demands of the motion planner are then handled by the framework which takes care of assigning or withdrawing resources from the algorithm based on the overall resource utilization.

4.1.1 Algorithm Design

State of the art RRT algorithms only deal with finding existing solutions as fast as possible. However, they do not take the problem's requirements into account for adapting the algorithm's resource usage. On the contrary, parallel RRT algorithms use a static number of processing units for calculation. They do neither consider resource availability nor the overall system state as explained in Section 2.7.3.

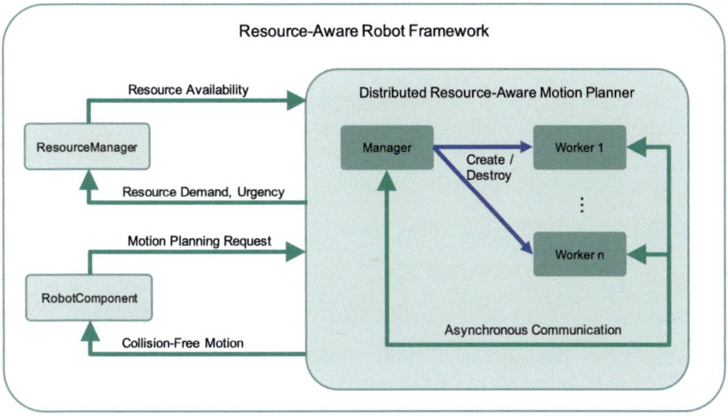

Figure 4.1: The motion planning algorithm consists of a manager process which creates and destroys functionally equivalent worker processes. Self-monitoring strategies are used to adaptively change the internal resource demand. Embedded within a resource-aware robot software framework, this algorithm is additionally capable of handing resources back to the framework if requested by a higher level resource manager.

The presented resource-aware RRT algorithm provides a first step into the resource-aware direction. It monitors its state and adapts its resource demand to the current planning problem. The design goals of this algorithm are:

- parallelism
- distributed computation
- low resource utilization for easy planning problems
- efficient resource utilization for difficult planning problems
- dynamic adaptation of parallelism to estimated planning problem difficulty
- capable of releasing resources on request
- system load reduction while still meeting QoS

Distributed computation and parallelism are achieved in the resource-aware motion planning algorithm by building on the Bulk Synchronous Distributed RRT which additionally provides reduced communication overhead as compared to other parallelized RRT implementations.

Local planners are based on a combination of Informed RRT* and Adaptive Dynamic-Domain RRT. The used algorithms are explained in more detail in Section 2.7. Informed RRT* was chosen due its asymptotic optimality and its improved convergence towards an optimal solution. Adaptive Dynamic-Domain RRT handles small passages in free-space better than standard RRTs by limiting its sampling domain to an approximation of the visible Voronoi region. Using other RRT implementations is also possible. However, some kind of failure metric must be provided by alternative planners in order to support higher-level resource allocation strategies which approximate the planning progress at runtime.

The planning algorithm itself consists of a single manager process and multiple functionally equivalent worker processes which execute the local planners as shown in Figure 4.1. Work associated with costly collision checks is parallelized instead of optimizing specific details of the algorithm such as sampling strategies whose computation time is comparatively low.

Algorithm 7 describes the manager process in detail. The main tasks of the manager are monitoring the planning state, analyzing the sampling tree, and determining if additional workers should be started based on resource allocation strategies. Sampling or collision checking is however not performed by the manager.

Algorithm 7 The actions performed by the manager process of the resource-aware motion planning algorithm.

Require: start and goal configurations $\rho_{start}, \rho_{goal} \in C_{space}$, allocation strategy s, bulk size $m \in \mathbb{N}$, planningTimeout, initial and maximum worker count workerCount$_{initial}$, workerCount$_{max}$, adaptive dynamic domain parameters α, $r_{border}, r_{min} \in \mathbb{R}$

procedure RESOURCE-AWARE_PLANNING_MANAGER
    $\tau \leftarrow$ NEW_SAMPLING_TREE()
    ADD_NODE(τ, ρ_{start})
    for $i \leftarrow 0..$workerCount$_{initial}$ **do**
        START_ADDITIONAL_WORKER()
    end for
    while ! (HAS_PATH(τ, ρ_{start}, ρ_{goal}) or IS_TIMEOUT()) **do**
        WAIT()
        APPLY_PENDING_UPDATES(τ)
        UPDATE_STRATEGY(s)
        if CURRENT_WORKER_COUNT() $<$ workerCount$_{max}$ **then**
            if REQUIRES_ADDITIONAL_RESOURCES(s) **then**
                START_ADDITIONAL_WORKER()
            end if
        end if
        while CURRENT_WORKER_COUNT() $>$ workerCount$_{max}$ **do**
            STOP_NEWEST_WORKER()
        end while
    end while
    SHUTDOWN_ALL_WORKERS()
    APPLY_PENDING_UPDATES(τ)
    return PATH(τ, ρ_{start}, ρ_{goal})
end procedure

Initially, the manager creates a sampling tree, adds the start node to the tree, and starts workerCount$_{initial}$ workers. Afterwards, the planning loop is executed until a path from ρ_{start} to ρ_{goal} is found or a timeout occurs after exceeding a predefined planning time.

The planning loop first waits until updates from worker nodes have arrived or a fixed time has passed (line 8). Next, updates from all workers are applied to the local search tree τ (line 9) and the internal state of the resource allocation strategy is updated (line 10). These resource allocation strategies are described in detail in Section 4.1.2. An additional worker is started (line 13) if the number of active workers is below workerCount$_{max}$ and if the allocation strategy determines that additional resources are required (`RequiresAdditionalResources(s)` evaluates to *true*). It must be noted that a newly started worker will obtain a private copy of τ first, before starting to plan.

The maximum number of active workers (workerCount$_{max}$) is designed to be adjusted from a higher level resource manager located outside the motion planner. Therefore, the algorithm reduces the active worker count (line 17) if it surpasses workerCount$_{max}$. The algorithm stops workers, starting with the newest worker, until the number of active workers is equal to workerCount$_{max}$.

After the planning loop finishes, all workers are shut down, remaining updates are applied to the search tree τ and the result is returned. The path from ρ_{start} to ρ_{goal} is returned on success and an empty path on failure. Planning fails if either no solution is found or if a timeout, specified before starting the algorithm, passes.

The overall computation time required by the manager is split mainly between two tasks. Starting worker processes and integrating their updates into the local search tree is the first task which usually does not require much processing power. Updating and evaluating the resource allocation strategies is the second important task whose processing power requirements depend on the selected strategy. If a cheap to evaluate allocation strategy is selected, the overall computation time of the manager process is negligible.

Algorithm 8 describes the local planner executed by each worker. The planner combines Informed RRT*, Adaptive Dynamic-Domain RRT, and Bulk Synchronous Distributed RRT.

Algorithm 8 Actions performed by each worker process of the resource-aware motion planning algorithm.

Require: goal configurations $\rho_{goal} \in C_{space}$, bulk size $m \in \mathbb{N}$, adaptive dynamic domain parameters α, r_{border}, $r_{min} \in \mathbb{R}$

procedure RESOURCE-AWARE_PLANNING_WORKER
  $\tau \leftarrow$ GET_CURRENT_TREE()
  while ! RECEIVED_SHUTDOWN_REQUEST() **do**
    for $i = 0..m$ **do**
      APPLY_PENDING_UPDATES(τ)
      repeat
        $\rho_{rnd} \leftarrow$ SAMPLE(C_{space})
        $\rho_{nearest} \leftarrow$ NEARESTNEIGHBOR(τ, ρ_{rnd})
      until $\rho_{nearest}.r >$ DISTANCE($\rho_{nearest}$, ρ_{rnd})
      $\rho_{reached} \leftarrow$ STEER($\rho_{nearest}$, ρ_{rnd})
      if $\rho_{reached} \neq \rho_{nearest}$ **then**
        $V_{NNs} \leftarrow$ NEARESTNEIGHBORS(τ, $\rho_{reached}$)
        $\rho_f \leftarrow$ SELECTPARENT_RRT*($\rho_{reached}$, $\rho_{nearest}$, V_{NNs})
        ADD_NODE(τ, $\rho_{reached}$)
        ADD_EDGE(τ, (ρ_f, $\rho_{reached}$))
        $\rho_{reached}.r \leftarrow \infty$
        $\rho_{nearest}.r \leftarrow \rho_{nearest}.r * (1 + \alpha)$
        REWIRE_RRT*(τ, $\rho_{reached}$, V_{NNs})
        UPDATE_MINIMAL_PATH_LENGTH(τ)
      else
        if $\rho_{nearest}.r = \infty$ **then**
          $\rho_{nearest}.r \leftarrow r_{border}$
        else
          $\rho_{nearest}.r \leftarrow \max(r_{min}, \rho_{nearest}.r * (1 - \alpha))$
        end if
      end if
    end for
    SENDUPDATE(τ.currentUpdate)
  end while
  SENDUPDATE(τ.currentUpdate)
end procedure

Upon startup, each worker fetches the current search tree from the manager and stores it in the local search tree τ. The planning loop executes until a shutdown request is received from the manager. Within the loop, m planning iterations are performed before broadcasting a bulk of m updates to the manager and all active planning nodes (line 28).

Each iteration starts with applying updates from all other workers to τ. Next, the *Dynamic Domain RRT* adaptation of the sampling domain is performed (line 6–9), followed by expanding towards ρ_{rnd} starting from $\rho_{nearest}$. If an intermediate solution $\rho_{reached} \neq \rho_{nearest}$ was found, a parent is selected (line 13), $\rho_{reached}$ is added to the search tree, and rewiring (line 18) is performed analogous to *RRT**. The radius of $\rho_{nearest}$ is increased analogous to *Adaptive Dynamic Domain RRT* (line 17). If no new configuration $\rho_{reached}$ could be found, the radius of $\rho_{nearest}$ is reduced (line 20–26). If a shutdown request is received, a final update of the current planning progress is sent as broadcast (line 30).

4.1.2 Resource Allocation Strategies

Different resource allocation strategies can be used to determine the necessity of additional computational power based on the difficulty and the progress of the current planning problem. These strategies are used during the initial planning phase for finding an initial solution. Resource allocation strategies play a secondary role after a solution was found, since the second planning phase optimizes the found solution and does not require to be computed in parallel. In order to allow replacing the local planner, each strategy must be independent of internals of the used planning algorithm. Additionally, a strategy should only cause minimal overhead and must therefore be computational inexpensive.

Approximation of the Planning Problem Difficulty

The goal of resource allocation strategies is to indicate if more resources are required based on the planning problem difficulty. Directly quantifying the difficulty would require finding small passages in the planning problem which requires building a representation of C_{free} and is therefore equivalent to solving the planning problem. Hence, the difficulty must be approximated incrementally based on the current state of the planner.

A first approach is to measure the overall computation time of the algorithm, since easy problems are usually solved in a short time frame and finding a solution for a more difficult planning problem will take much longer. This metric is therefore very robust since a planning problem is guaranteed to be difficult if it takes a long time to solve. Furthermore, this metric is usable with any type of planner, since it is independent of the planning algorithm. Consequently, the idea is to measure the algorithms runtime and request more resources after a specific time ΔT has passed since additional resources were requested. However, the correct choice of ΔT determines if additional workers are started too frequently (ΔT too small) or too infrequently (ΔT too large). This problem can be addressed by adding an additional parameter to compensate for bad choices of ΔT, similar to how bad choices of r_{border} are addressed by Adaptive Dynamic-Domain RRT. Overall, weaknesses of other approaches can always be balanced out by adding a time delta to the strategy.

A second approach requires the underlying algorithm to provide some kind of failure metric. One possibility is to determine the ratio between newly created configurations and the total amount of attempts for creating new configurations over the last N iterations. In case of Dynamic-Domain RRT and related algorithms this ratio provides an approximation of the boundary surface between the tree region and the tree region $\cup C_{obstacle}$. A high ratio indicates that the problem is most probably hard, since the algorithm is currently searching near the boundary of an obstacle. In case of a high ratio new resources can be requested. However, it is essential for the planning algorithm to support a similar failure metric, since this ratio tends to be relatively high if only random sampling is used.

Other approaches can track the expansion of the search tree and us this metric to calculate the ratio between current expansion and maximum expansion. The first kind of expansion metric depends on the cost or length of a path. Expansion can be detected by monitoring the average cost for reaching a leaf node or the length of the longest path in the tree. An estimate for the current expansion rate is provided by observing the length of the last N edges which were added to the search tree. Major drawbacks of using path length or costs as metric is the necessity of a reference value in order to correctly associate and compare the length or cost to the peculiarities of the configuration space. Another kind of expansion metric considers the volume of the search tree. One approach is to use the volume of the Axis Aligned Bounding Box (AABB) of all generated nodes and compare it to the overall volume of the configuration space. Similar to the previous length and cost related metrics, the AABB version suffers from the same drawback regarding reference values. Additionally, for this metric to be considered stable, the AABB volume must be guaranteed to expand only if the solution is approached. This is however not possible for cases where the solution is contained within the AABB's volume in which case it is impossible to track the planning progress.

Implemented Resource Allocation Strategies

After evaluating the presented approaches for approximating the planning problem difficulty, two of them were selected and implemented as resource allocation strategies. The output of both strategies is a binary *true* or *false* indicating whether additional resources should be requested or not. Distance-based metrics were not implemented, since they depend on the planning problem and would require normalizing the input. However, normalizing the input is not possible in a general, meaningful, and problem independent way.

The strategy implemented first makes use of the ΔT-approach and is referred to as ΔT-strategy and as ΔTy for a parameter $\Delta T = y$. This strategy was an obvious choice, since planning time correlates directly with the problem difficulty. Equation (4.1) provides the basis for deciding if additional resources are required. The equation evaluates to *true* for the current time T_{now}, if more than ΔT time has passed since the last time T_{last} when additional resources were requested.

$$T_{last} + \Delta T \leq T_{now} \tag{4.1}$$

Choosing a small value for ΔT leads to frequent resource requests while high values of ΔT leads to belated resource acquisition. While the first case leads to underutilization of resources, the second case leads to delays in planning time.

The second strategy additionally modifies the ΔT parameter according to the second approach presented in Section 4.1.2, to overcome this problem. ΔT is modified according to the rate φ of failed node creations versus the total number of node creation attempts. For high ratios of φ, the algorithm is most likely searching near difficult obstacle boundaries and would benefit from additional resources. An additional parameter σ is introduced which is used to dampen the effect of φ on ΔT. This strategy is referred to as *NN*ΔT-strategy (No Node ΔT) and as *NNx*ΔTy for a set of parameters $\sigma = x$ and $\Delta T = y$.

$$\Delta T' = \frac{\Delta T}{1 + \sigma * \varphi} \tag{4.2}$$

$$T_{last} + \Delta T' \leq T_{now} \tag{4.3}$$

Equation (4.2) describes how σ and φ influence the ΔT parameter and equation (4.3) describe the formula based on which the *NN*ΔT-strategy decides to request additional resources. Overall, the *NN*ΔT-strategy requests additional resources at least every ΔT but not faster than every $\Delta T/(1+\sigma)$.

Figure 4.2 shows the influence of σ on ΔT based on the value of φ (failed node creation attempts). The faster the curve descends, the smaller the time $\Delta T'$ after which new resources are requested. A value of $\sigma = 0$ results in the regular ΔT strategy, while increasing values of σ lead to shorter acquisition intervals. This means that the *NN*ΔT-strategy can compensate for high values of ΔT for planning problems by requiring more resources in a faster time interval when the planner is searching in a narrow passage.

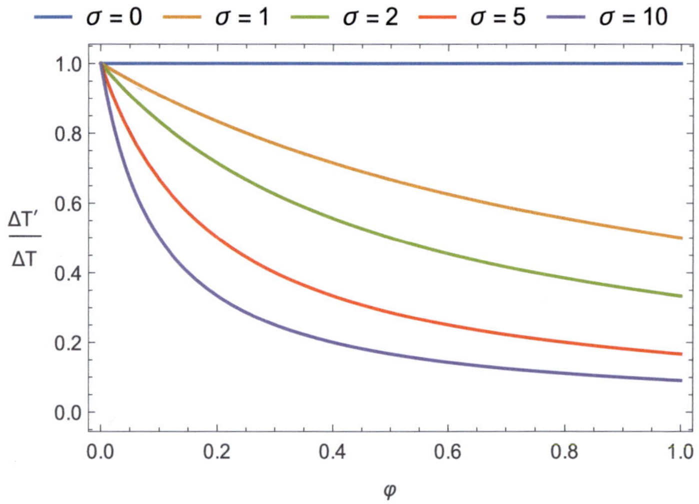

Figure 4.2: Effect of the parameter σ of the *NN*ΔT-strategy on the ratio $\frac{\Delta T'}{\Delta T}$ as a function of φ. Higher values of σ lead to shorter $\Delta T'$ times. *Source:* [Kröhnert et al., 2016] ©2016 IEEE

4.1.3 ArmarX Integration

The presented resource-aware and distributed planning algorithm is implemented as part of the ArmarX robot framework. A more detailed description of ArmarX can be found in Appendix C. Communication between ArmarX components is either performed via Remote Procedure Calls (RPC) or a Publish/Subscribe mechanism. The architecture of the planning algorithm builds on these foundations and consists of the following four major parts:

1. RemoteObjects provide computational resources by starting threads remotely. They allow communicating via RPC or broadcast messages.

2. RemoteObjectNodes represent computational resources and provide the mechanisms for starting and stopping RemoteObjects. A RemoteObjectNode must be started on each PC in the distributed network which is supposed to provide computational resources.

3. PlanningTasks contain all relevant information of a motion planning problem such as ρ_{start}, ρ_{goal}, the specific planning algorithm to use, and information about the size of the C_{space} and the files required for performing collision checks. Additionally, the PlanningTask implements the resource allocation strategies in order to provide them to the dedicated planning algorithms.

4. PlanningServer is the central component which receives PlanningTasks and executes them sequentially. The PlanningServer also holds a list of RemoteObjectNodes used by the manager of the motion planning algorithm to acquire new computational resources.

Initially, a client creates and parameterizes an instance of PlanningTask, sends it to the PlanningServer, and receives a proxy. This proxy allows querying the now server-side PlanningTask for the planning progress and the final planning result via Remote Procedure Calls. The PlanningServer adds the

PlanningTask to its task queue and starts it once all previous tasks have finished computing their motions. All available RemoteObjectNodes can be used by the PlanningTask to start and stop remote computations by creating and destroying RemoteObject instances on these nodes. After planning finishes, the used RemoteObjects are destroyed, the solution is stored in the PlanningTask, and control is handed back to the PlanningServer for executing the next task. Figure 4.3 shows a sequence diagram of the previously described steps involved in planning a motion.

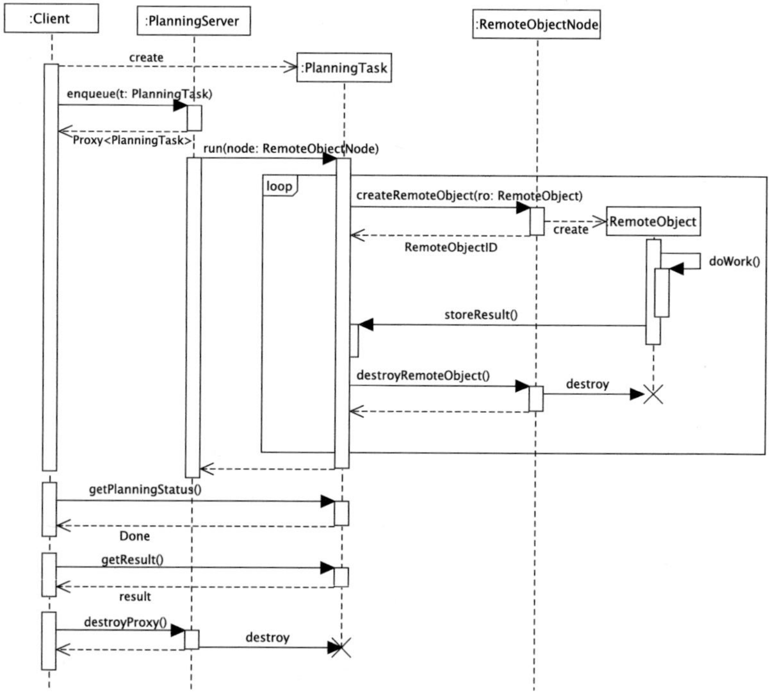

Figure 4.3: Sequence diagram of the steps involved from specifying a planning problem to retrieving the result.

4.2 Resource-Aware Disparity Map

Disparity maps are two dimensional images whose pixel colors correlate to the distance of objects visible in the image. Light gray pixels are close to the viewer while darker gray pixels represent objects with a larger distance. Disparity maps can be computed from a pair of stereo images captured by calibrated stereo cameras. The algorithm for computing the disparity map used in this work was published in [Faugeras et al., 1993]. Computing the disparity map is computationally expensive. However, the algorithm offers a high degree of parallelism, since it works on isolated regions of the image.

Parallelized algorithms are typically mapped to resources in a static manner. To overcome inefficiencies of static resource mapping, new resource-aware methodologies such as Invasive Computing [Teich et al., 2011] are currently studied. In this section a parallelized version of the disparity map algorithm is shown as an example of invasive algorithms. The disparity map algorithm is capable of benefiting from resource-aware concepts such as online algorithm adaptation. This is achieved through dynamic problem partitioning and resource allocation based on the currently available resources.

In the following, two different implementations of the resource-aware disparity map algorithm are presented. Both versions are designed to run on an invasive platform with multiple processors. The first implementation is written in the X10 programming language [Charles et al., 2005]. X10 was adopted and extended by InvasIC to provide all required features related to resource-awareness. The second version is written in C++ and interfaces directly with the invasive operating system named OctoPOS [Oechslein et al., 2011]. This approach enables fast algorithm development using an emulation of OctoPOS while being able to execute the application on physical invasive hardware by simply rebuilding the application for the new target.

4.2.1 Invasive X10 Implementation

The X10 implementation of the disparity map algorithm is available both as a single-core as well as a resource-aware multi-core version. Resource-aware features missing in X10 are provided by the Invasive X10 library and a functional simulation tool [Hannig et al., 2011]. A reference implementation of a disparity map algorithm in C++ is provided by the Integrating Vision Toolkit (IVT) [1] [Azad, 2009]. Calculating the disparity information from a pair of stereo input images is performed by moving a sliding window simultaneously across both images. The reference implementation in IVT processes input images in a single large but efficient loop by reusing previously computed results.

The single-core X10 version of the disparity map algorithm resembles the IVT implementation. However, the X10 implementation does not take advantage of any resource-aware concepts and is tailored to computer architectures as they are found in current humanoid robots such as ARMAR-III. Typical commonalities of these hardware architectures are large amounts of main memory accessible by the processor via a fast interconnect.

On the contrary, embedded systems, typically containing less main memory but potentially more CPU cores, would not be able to execute this algorithm directly, at least not very efficiently. Due to higher resource constraints, such embedded systems benefit from resource-aware approaches which provide algorithms with the means to adapt to these constraints. Therefore, a resource-aware multi-core version of the disparity map algorithm was implemented in Invasive X10 which is capable of adapting the internal parallelism based on the quantity of available resources.

Achieving resource-awareness in algorithms requires starting with a parallelized version of the algorithm and enhancing it to react to resource changes and requirements at specific points during execution. Furthermore, these algorithms must be executed on a multi-core or many-core architecture in order

[1] `http://ivt.sourceforge.net/`

to take advantage of their parallel nature. The architecture used for executing and evaluating the invasive disparity map contains a shared main memory and 8 RISC processors also called Processing Elements (PEs), see Figure 4.4. The processors are divided into two groups called *tile*s each containing 4 processors and a block of local shared memory called tile-local memory (TLM). Both tiles and the main memory are connected via a Network on Chip (NoC). Thus, access to TLM is faster than to the main memory because it can be accessed locally and does not have to fetch data over the much slower Network on Chip connection.

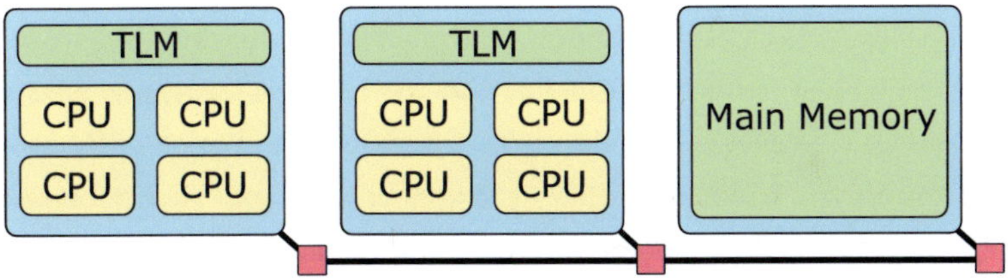

Figure 4.4: The hardware layout used for executing the resource-aware disparity map algorithm. A total number of 8 CPUs is distributed across two tiles containing additional Tile-Local Memory (TLM). Both compute tiles and a third tile containing the main memory are connected via a Network on Chip (NoC).

Taking the execution platform into account, it is recognizable that an algorithm such as the single-core disparity map which manipulates data stored in main memory will take a long time to execute. This is due to the constant main memory accesses putting heavy load on the NoC. By definition, accessing the NoC is slower than accessing local memory partitions such as the TLM.

Getting the best performance out of the underlying many-core architecture requires building a parallelized distributed version of the algorithm and enhancing it with resource-aware features. The disparity map algorithm is therefore modified to split up the computation into fine-granular work packages. Every work package calculates a partial result which is merged into the final result image upon completion. Processing the smaller work packages is performed by the same basic implementation as it is used in the single-core implementation. The corresponding high-level coordination of the parallelized algorithm follows a divide & conquer approach and is shown in Algorithm 9.

After receiving input images, the algorithm starts by requesting the maximum number of PEs specified as input parameter maxPEs (line 2). Available resources returned by the invasive platform are stored in a structure called claim which can be queried for a valid set of resources (line 3).

Dynamic and adaptable splitting of the pixel data of the input images into smaller blocks is performed from line 4 to line 12. First, the total number of usable PEs is extracted from the claim, followed by a division of the input data into equally sized blocks (line 8). It is essential to split the input images into overlapping blocks in order to produce an output image without missing lines. A sliding window of size n reduces all pixels covered by the window into a single pixel. Thus, every block but the first has to have at least $n-1$ lines overlap in order to generate a complete output image. Furthermore, both PE count as well as the amount of available TLM are taken into account for determining the size of the image blocks. This allows the algorithm to run on systems equipped with only a small amount of TLM.

The split image data is distributed to the TLMs of the tiles on which the acquired PEs are located. A call to `infect()` then distributes the worker functions to the respective PEs and starts the computation. Afterwards, the coordination function waits for the workers to finish their computations (line 16) before the different parts of the result image are collected and combined into the final disparity map.

Once all calculations are finished and their results collected, all requested resources are released again by calling `retreat()` on the previously acquired *claim* (line 22).

In contrast to algorithms using static resource allocation, this resource-aware version is capable of calculating a disparity map even if the amount of available resources is constantly changing. If the request for additional resources can only be partially fulfilled during execution, the algorithm can react in two different ways. With enough tile-local memory but less PEs available, the image chunk size is increased. If both less memory and less PEs are available, the amount of image chunks is increased while keeping the chunk size small.

4.2.2 Invasive C++ Implementation

The invasive X10 version of the disparity map algorithm was also implemented as a native C++ application directly on top of OctoPOS, the invasive operating system. This approach allows running the algorithm on real invasive prototype hardware and analyzing the benefits of using tile-local memory over regular main memory. Figure 4.5 shows the used hardware platform, consisting of a single tile containing 6 CPUs running at 80 MHz.

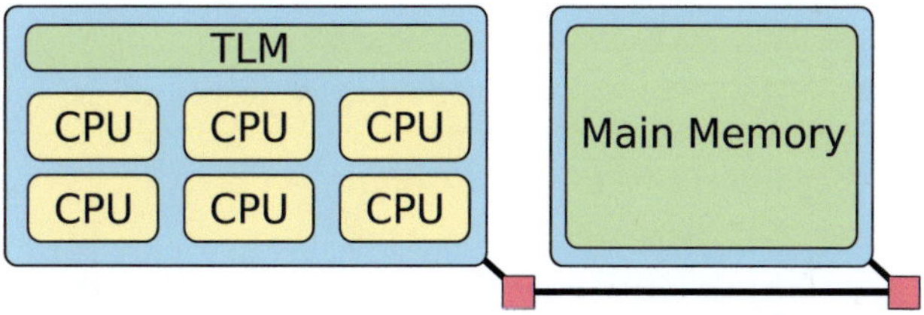

Figure 4.5: The hardware layout used for executing and evaluating the resource-aware disparity map algorithm in C++. A total number of 6 CPUs are grouped together in one compute tile containing additional Tile-Local Memory (TLM). The compute tile as well as the main memory tile are connected via a Network on Chip (NoC).

Algorithm 9 Coordination function of the resource-aware disparity map algorithm.

```
Require: maxPEs, inputImages
procedure DISPARITY_MAP_COORDINATION
    claim ← invade(maxPEs)
    if claim.valid() then
        n ← claim.PEs().count()
        // retrieve the list of available memory for each processor
        localMemory[] = claim.getLocalMemorySize()
        // divide the data according to available resources and memory
        data[] ← divideIntoBlocks(inputImages.pixelsLeft,
                        inputImages.pixelsRight, n, localMemory)
        // distribute data blocks to the PE's local memory
        for pe ∈ claim.PEs() do
            pe.setLocalMemory(data[pe.id])
        end for
        // distribute the disparity map code onto the PEs and start it
        infect(localDisparityMapComputation())
        // wait for all PEs to finish their computation
        waitForParallelCalculations()
        // collect results in disparityMap
        for pe ∈ claim.PEs() do
            disparityMap.pixels[pe.id] ← pe.getLocalMemory()
        end for
        // free unused resources
        retreat(claim)
    end if
    return disparityMap
end procedure
```

4.2.3 Invasive Computing Head Demo

To integration of the disparity map into a bigger context was demonstrated in a setup using the Karlsruhe Humanoid Head [Asfour et al., 2008]. The task of the robot head is to track a colored object, once it is recognized in the captured camera images. An image of the setup is shown in Figure 4.6. Initially no target object is assumed to be present in the scene, thus most resources are dedicated to calculating a disparity map image. Finding skin and object colored regions is performed on low resolution images with a minimal set of resources. Once, the object is detected in the low-resolution images, tracking of the object starts and most resources are dedicated to calculating high quality object positions. Once the object is lost, the object recognition falls back to using low-resolution images and the majority of the processing power is again used for calculating the disparity map.

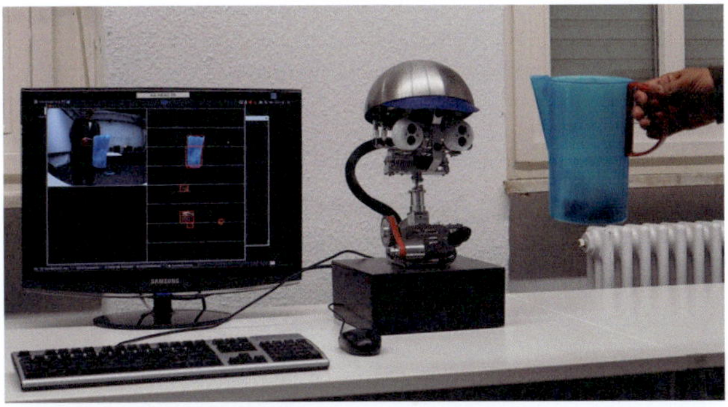

Figure 4.6: The standalone head of the humanoid robot ARMAR-III tracking a blue pitcher. The object position is determined by an invasive color segmentation algorithm.

The robot head is controlled by the original head software. Stereo images are forwarded to the invasive C++ algorithms running on the OctoPOS guest layer on the same PC. Using the OctoPOS guest layer was the option chosen to provide the head control software with fast image updates, which are required to show online tracking capabilities. Due to the lower processing speed of 40 MHz of the processors available on the invasive prototype hardware the tracking frequency would have been reduced significantly. The layout of the used invasive architecture is shown in Figure 4.7. It contains 8 CPU cores divided among 2 compute tiles, one main memory tile and one I/O tile. The I/O tile is used to receive and transmit images and other data from within OctoPOS.

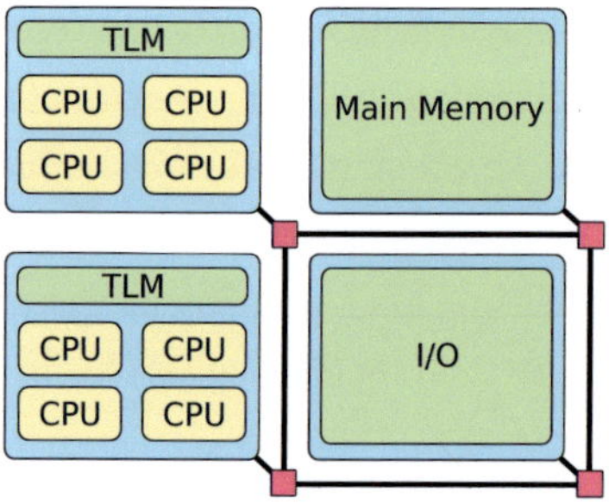

Figure 4.7: The hardware layout used for running the invasive head demo OctoPOS guest layer. A total number of 8 CPUs is distributed across 2 tiles containing additional Tile-Local Memory (TLM). All compute tiles as well as a memory tile and an I/O tile are connected via a Network on Chip (NoC).

5 Evaluation

The following different aspects of this work are evaluated in this chapter. First, examples of resource models are presented. Next, the prediction of future robot tasks and their associated resource utilization are evaluated in a simulated *pick and place* scenario. Last, results of the resource-aware motion planner and the invasive disparity map algorithm are shown.

5.1 Profiling and Resource Models

Resource models generated from profiling data are shown in this section. The profiling data was obtained from a *pick and place* task executed by the humanoid robot ARMAR-IIIa in simulation. Figure 5.1 demonstrates the simulated task execution with screenshots while the executed statecharts are shown in Figure 5.2 and Figure 5.3. Transition data was stored upon occurrence of the associated transition events. CPU and memory utilization data was reported and stored periodically every 300 ms. The execution platform used for evaluation was defined by the following platform model $P := (3.40\,GHz, 8, 16\,GB\,RAM, 1\,GBit)$. This specification is close to the one of the ARMAR-IIIa coordination PC. Important parameters contained in the context were the pick up location (sideboard), the place location (table), the object to move (cereal box), but no further obstacles or humans.

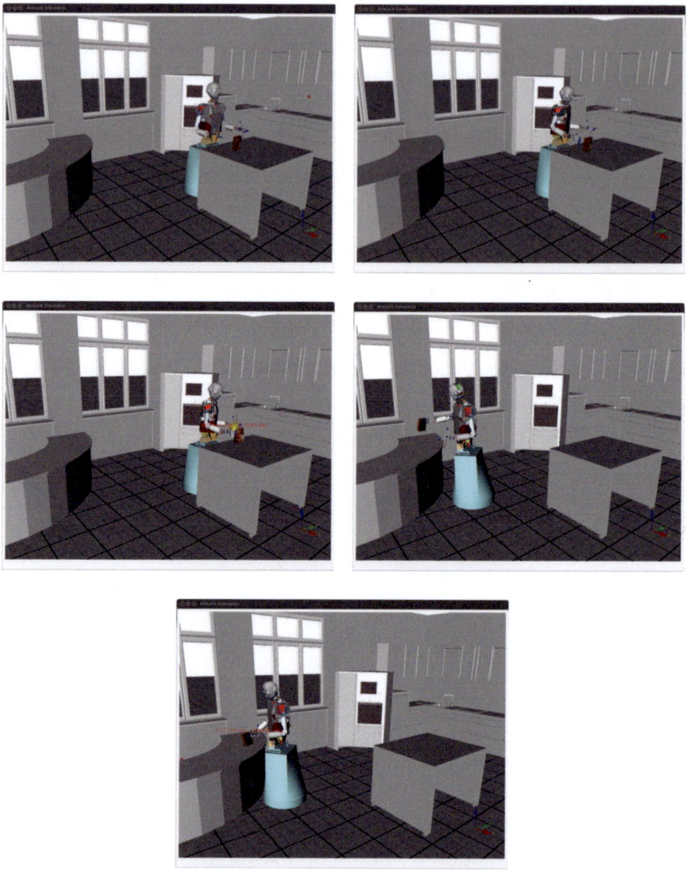

Figure 5.1: ARMAR-IIIa first moves from its initial position to the sideboard (top left) tries to detect the requested object and moves its hand to the object if it is found (top right). After ARMAR-IIIa reached for the object and grasped it, the next step is to lift the object (middle left), followed by moving to the table (middle right), and finally placing the object on the table (bottom).

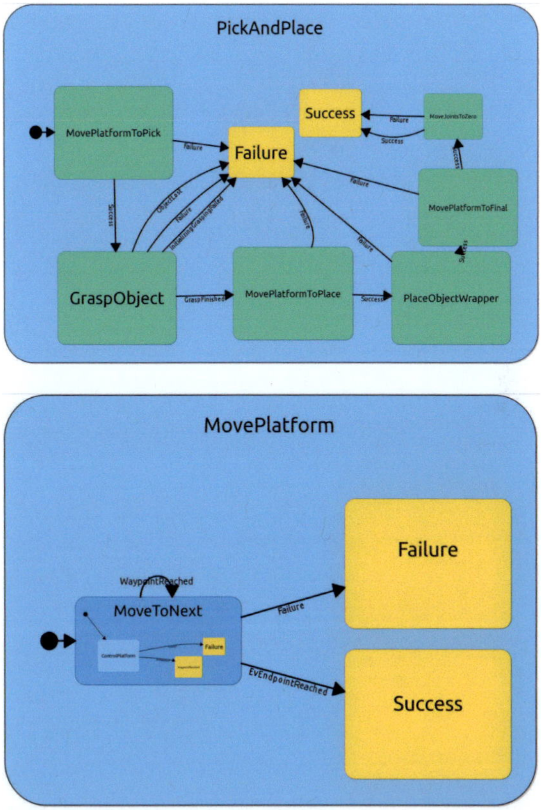

Figure 5.2: The *pick and place* statechart at the top displays the different required robot tasks. States marked in yellow are end states and states marked in green are statecharts which are executed in different processes. If an end state is entered, it sends a transition event and immediately exits its parent state. The bottom statechart shows the *move platform* statechart used for driving the robot through an environment.

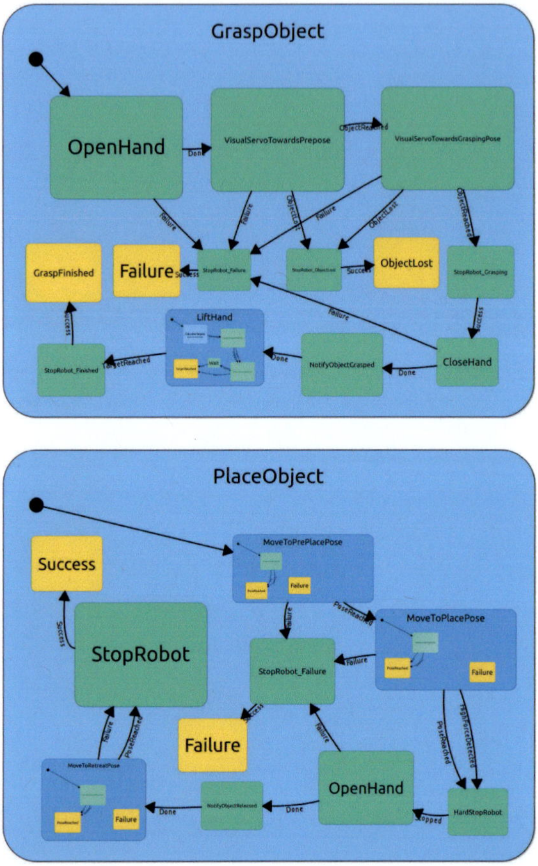

Figure 5.3: Second part of Figure 5.2 containing the *grasp object* and *place object* statecharts. *grasp object* opens the robot hand, uses visual servo to reach the object, grasps the object, and finally lifts the object. The statechart contains additional states for handling different kind of errors which can occur during visual servoing or object localization. The *place object* statechart moves the robot arm to the placing position, opens the hand if a contact between object and surface was detected, and retreats the arm.

The transition profiling data obtained from the execution is depicted in Figure 5.4 showing the execution intervals/durations of all states involved in the task execution. Each line in the state execution diagram represents the duration of a specific state with an index. The state name matching the index can be retrieved from the mapping presented in Table 5.1. Each $\rightarrow$ in the state name indicates a sub-state relation with the top-level parent on the left and the executing state on the right side of the rightmost $\rightarrow$. Every state execution is colored differently while the duration is represented by the length of the respective bar. Additionally, states with shorter duration are sub-states of states with longer duration. The topmost *pick and place* task for example has index 50 and can be found at the top of the figure.

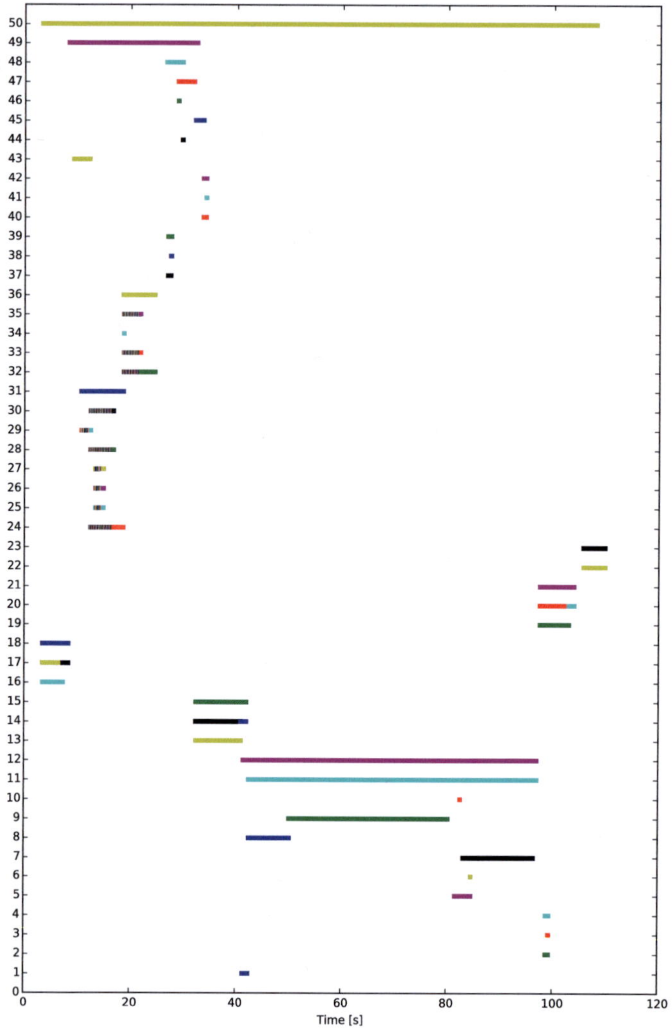

Figure 5.4: Execution intervals (durations) of the states involved for executing a pick and place task. A state X whose execution interval is contained in the interval of another state Y indicates that X is a sub-state in the hierarchy below state Y. The mapping between indices and state names can be found in Table 5.1.

Table 5.1: The mapping between indices and state names for Figure 5.4. Each line represents the state hierarchy starting with the top-level on the left and → pointing to the next sub-state in the hierarchy. The current state's name is thus the name after the last → of each entry.

Index	State Name
50	PickAndPlace
49	PickAndPlace→GraspObject
48	PickAndPlace→GraspObject→CloseHand
47	PickAndPlace→GraspObject→LiftHand
46	PickAndPlace→GraspObject→LiftHand→CalculateTargets
45	PickAndPlace→GraspObject→LiftHand→Wait
44	PickAndPlace→GraspObject→NotifyObjectGrasped
43	PickAndPlace→GraspObject→OpenHand
42	PickAndPlace→GraspObject→StopRobot_Finished
41	PickAndPlace→GraspObject→StopRobot_Finished→MoveJoints
40	PickAndPlace→GraspObject→StopRobot_Finished→SetZeroVelocity

Table 5.1: The mapping between indices and state names for Figure 5.4. Each line represents the state hierarchy starting with the top-level on the left and → pointing to the next sub-state in the hierarchy. The current state's name is thus the name after the last → of each entry.

Index	State Name
39	PickAndPlace→GraspObject→StopRobot_Grasping
38	PickAndPlace→GraspObject→StopRobot_Grasping→MoveJoints
37	PickAndPlace→GraspObject→StopRobot_Grasping→SetZeroVelocity
36	PickAndPlace→GraspObject→VisualServoTowardsGraspingPose
35	PickAndPlace→GraspObject→VisualServoTowardsGraspingPose→CalculateTcpTarget
34	PickAndPlace→GraspObject→VisualServoTowardsGraspingPose→GetObjectInstance
33	PickAndPlace→GraspObject→VisualServoTowardsGraspingPose→GetObjectPose
32	PickAndPlace→GraspObject→VisualServoTowardsGraspingPose→VisualServoTowardsTargetPoseWrapper
31	PickAndPlace→GraspObject→VisualServoTowardsPrepose
30	PickAndPlace→GraspObject→VisualServoTowardsPrepose→CalculateTcpTarget
29	PickAndPlace→GraspObject→VisualServoTowardsPrepose→GetObjectInstance

Table 5.1: The mapping between indices and state names for Figure 5.4. Each line represents the state hierarchy starting with the top-level on the left and → pointing to the next sub-state in the hierarchy. The current state's name is thus the name after the last → of each entry.

Index	State Name
28	PickAndPlace→GraspObject→VisualServoTowardsPrepose→GetObjectPose
27	PickAndPlace→GraspObject→VisualServoTowardsPrepose→StopRobot
26	PickAndPlace→GraspObject→VisualServoTowardsPrepose→StopRobot→MoveJoints
25	PickAndPlace→GraspObject→VisualServoTowardsPrepose→StopRobot→SetZeroVelocity
24	PickAndPlace→GraspObject→VisualServoTowardsPrepose→VisualServoTowardsTargetPoseWrapper
23	PickAndPlace→MoveJointsToZero
22	PickAndPlace→MoveJointsToZero→MoveJoints
21	PickAndPlace→MovePlatformToFinal
20	PickAndPlace→MovePlatformToFinal→MoveToNext
19	PickAndPlace→MovePlatformToFinal→MoveToNext→ControlPlatform
18	PickAndPlace→MovePlatformToPick

Table 5.1: The mapping between indices and state names for Figure 5.4. Each line represents the state hierarchy starting with the top-level on the left and → pointing to the next sub-state in the hierarchy. The current state's name is thus the name after the last → of each entry.

Index	State Name
17	PickAndPlace→MovePlatformToPick→MoveToNext
16	PickAndPlace→MovePlatformToPick→MoveToNext→ControlPlatform
15	PickAndPlace→MovePlatformToPlace
14	PickAndPlace→MovePlatformToPlace→MoveToNext
13	PickAndPlace→MovePlatformToPlace→MoveToNext→ControlPlatform
12	PickAndPlace→PlaceObjectWrapper
11	PickAndPlace→PlaceObjectWrapper→PlaceObject
10	PickAndPlace→PlaceObjectWrapper→PlaceObject→HardStopRobot
9	PickAndPlace→PlaceObjectWrapper→PlaceObject→MoveToPlacePose
8	PickAndPlace→PlaceObjectWrapper→PlaceObject→MoveToPrePlacePose
7	PickAndPlace→PlaceObjectWrapper→PlaceObject→MoveToRetreatPose

Table 5.1: The mapping between indices and state names for Figure 5.4. Each line represents the state hierarchy starting with the top-level on the left and → pointing to the next sub-state in the hierarchy. The current state's name is thus the name after the last → of each entry.

Index	State Name
6	PickAndPlace→PlaceObjectWrapper→PlaceObject→NotifyObjectReleased
5	PickAndPlace→PlaceObjectWrapper→PlaceObject→OpenHand
4	PickAndPlace→PlaceObjectWrapper→PlaceObject→StopRobot
3	PickAndPlace→PlaceObjectWrapper→PlaceObject→StopRobot→MoveJoints
2	PickAndPlace→PlaceObjectWrapper→PlaceObject→StopRobot→SetZeroVelocity
1	PickAndPlace→PlaceObjectWrapper→WaitForHandAndObjectChannel

Figure 5.5 shows the CPU utilization of all robot components required during the execution of the *pick and place* statechart. Figure 5.4 shows the statechart execution characteristics in detail. Each robot component's CPU utilization is shown on a separate line starting with the robot component name postfixed with _CPU. CPU utilization for each component ranges from 0 % to 100 % within an area limited by the current and the next components' name. Most components show an extremely low CPU utilization between 0 % and 5 % such as HandMarkerLocalization (red line) and TCPControlUnit (green line) which are shown at the top of Figure 5.5. TCPControlUnit shows a small spike in the beginning and low activity during later timesteps, whereas HandMarkerLocalization shows almost no activity and stays near 0 % CPU utilization.

Some components like XMLStateComponentHandGroupApp drop back to a constant 0 % to 5 % after an initial high utilization of about 100 % while ViewSelection alternates between high and low utilization. The RobotStateComponent which is responsible for holding the current sensor values of the robot, utilizes around 5-10 % at the beginning, switches to 75-80 % after approximately 30 seconds, and then drops back to about 65-70 % after approximately 55 seconds. XMLStateComponentVisualServoGroupApp responsible for performing visual servoing shows phases with almost 0 % utilization when inactive and phases with approximately 50 % utilization when visual servoing is performed.

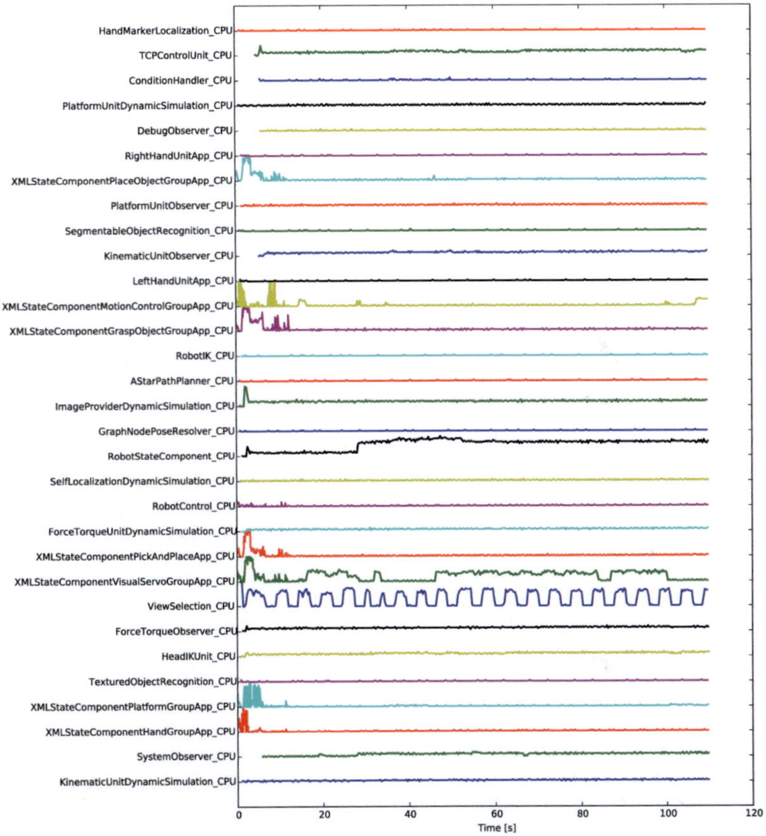

Figure 5.5: CPU utilization of running robot components.

Figure 5.6 shows the utilization of dynamic memory in MByte of 9 selected robot components. Components with low changes in memory utilization were excluded for better visibility.

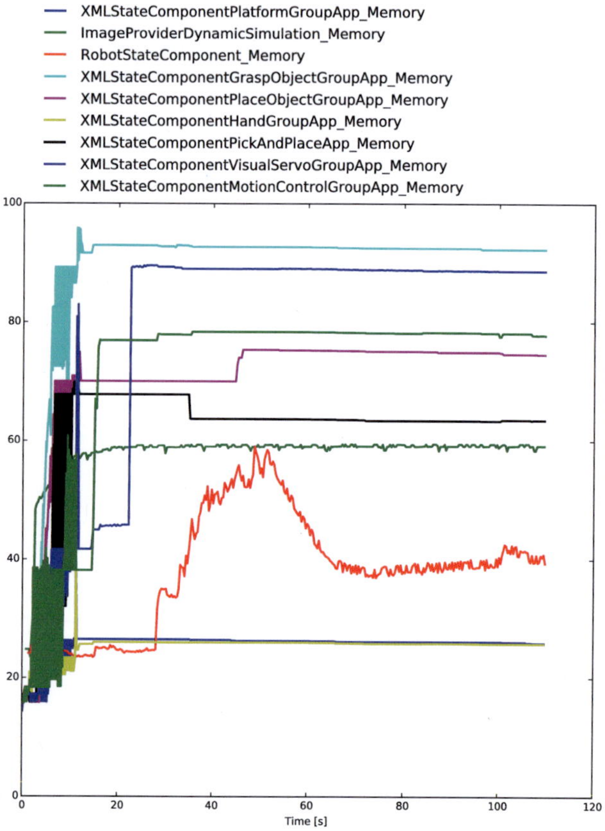

Figure 5.6: Memory utilization in MByte of a selection of robot components.

An initial fluctuation in allocation and deallocation of memory can be observed. The upper blue line in Figure 5.6 represents the XMLStateComponentVisualServoGroupApp whose memory allocation increases drastically after approximately 20 seconds, which is the time when the visual servoing statechart gets triggered. The red line represents the RobotStateComponent and shows a steady increase in memory allocation once the visual servoing starts and decreases, once the platform starts moving to the target position.

CPU utilization associated with the resource model of the transition ending in Place-Object is depicted in Figure 5.7 while the memory utilization of the resource model is shown in Figure 5.8. The PlaceObject sub statechart was started after 42 seconds and had a duration of 54 seconds.

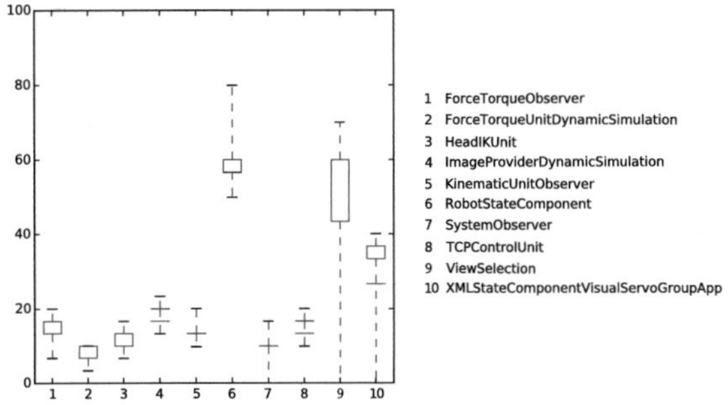

Figure 5.7: CPU utilization in [%] of robot components running during the execution of the PlaceObject state starting after 42 seconds.

The figure shows changes in CPU utilization of HeadIK, ViewSelection and RobotStateComponent. HeadIK is calculates the kinematics of the head and where the robot should look at. ForceTorqueUnit and ForceTorqueObserver perform processing of the force values in the wrist of the robot which is required for detecting contact with the table. SystemObserver, KinematicUnitObserver, and XMLState-ComponentVisualServoGroup also use CPU and are required for performing visual servoing of the hand to reach the target position.

Figure 5.8 shows that only a few components perform a significant amount of dynamic memory allocation in the PlaceObject state. The most significant change happens in RobotControlUnit which continuously updates the internal robot representation by applying the latest sensor updates and in the TCPControlUnit which is used to steer the robot's arm towards the target pose. Additional memory is allocated and deallocated in XMLStateComponentPlaceObjectGroup, the overall statechart component coordinating the subtask.

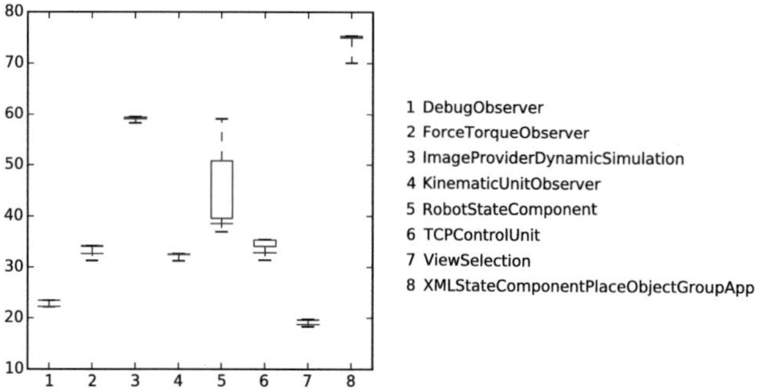

Figure 5.8: Memory utilization in [MB] of robot components running during the execution of the PlaceObject state starting after 42 seconds.

MoveToPrePlacePose is a sub-state of PlaceObject starting at 42 seconds with a duration of 7.6 seconds. The CPU utilization of the resource model of MoveToPrePlacePose is shown in Figure 5.9. Compared to the parent state, ViewSelection and XMLState-ComponentVisualServoGroup show wider variation in CPU utilization, while RobotStateComponent shows smaller variation in CPU utilization.

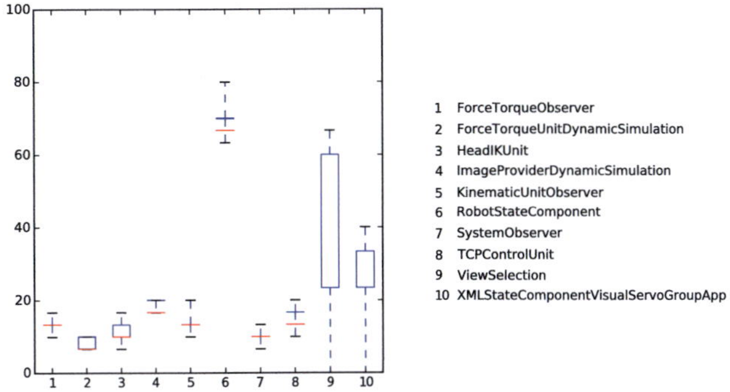

Figure 5.9: CPU utilization in [%] of robot components running during the execution of the MoveToPrePlacePose state which started after 42 seconds.

Memory utilization of the resource model of MoveToPrePlacePose is shown in Figure 5.10. The figure shows that a reduced number of components perform dynamic allocation during this step as compared to the overall utilization of the parent state. RobotStateComponent allocates a part of the memory which is found in the statistics of the parent state. To the contrary, the memory allocation statistics of XMLState-ComponentVisualServoGroup match the ones found in the parent state, indicating that this state is mostly responsible for the allocation.

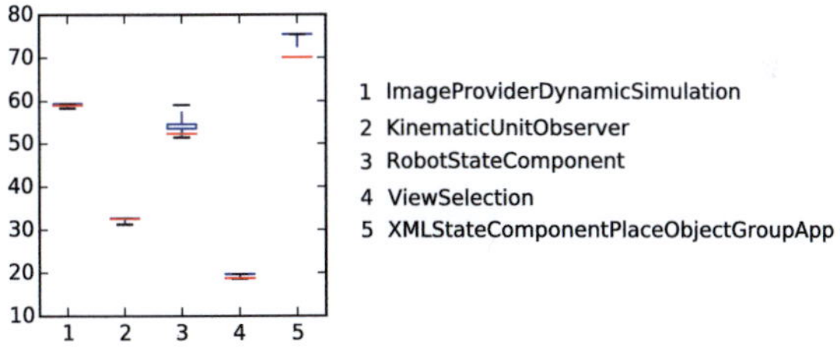

Figure 5.10: Memory utilization in [MB] of robot components running during the execution of the MoveToPrePlacePose state which started after 42 seconds.

5.2 Robot State and Resource Prediction

Evaluation of the resource prediction approach is performed in the context of a *pick and place* robot program implemented in ArmarX. The robot program describes a task for picking up an object from a table and placing it on another location such as the sideboard. The execution of this task contains the following subtasks: go to table, locate object, reach object, grasp and lift object, go to destination location, place object, release object, and retreat the arm. Irregularities in the execution of the task may occur, once a human operator performs one of the following tasks:

- enter the room and observe the robot from a distance (high probability)
- walk to the table and grasp an object (possibly the object the robot is supposed to grasp)
- leave the room

The human and all possible human tasks are described in and simulated by the statechart shown in Figure 5.11. Transitions between human tasks are annotated with probabilities to describe the likelihood of changing from one task to another one.

Once a person is present in the scene, it may intend to grasp an object located in front of the robot. The robot is required to react on such interruptions from the person to prevent any kind of damage. Possible reactions are aborting the current task and start a dialog with the person, or pausing and continue grasping the target object if the object grasped by the person is outside the manipulation space of the robot. These interactions are stored as part of the robot's context in MemoryX. Once a person performing similar tasks was encountered multiple times, it will influence the outcome of the prediction in future calculations.

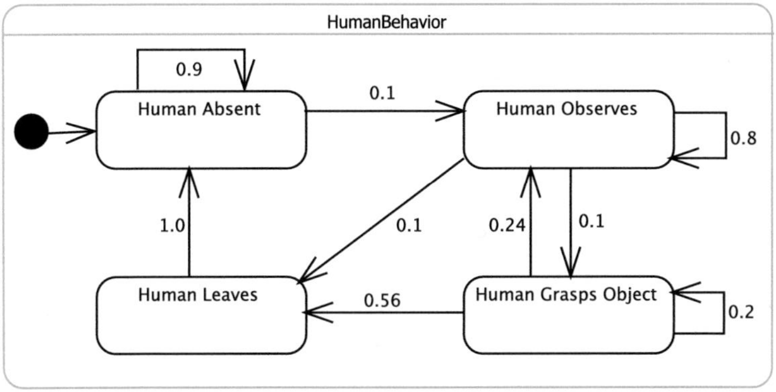

Figure 5.11: The behavior of a human regarding the workspace of the robot. The human can be absent, enter the scene and start observing the robot, grasp an object, or leave the scene. Transitions between the tasks are annotated with transition probabilities.

Figure 5.12 shows the complete statechart of the Pick And Place task including all error handling states. The entry point of the robot task is the *idle* state in which the robot receives a pick and place command for a specific object. The complete task is subdivided into a *pick task* and a *place task*. Once the location of the target object is known, the *pick task* is entered and continued with the *move to object location* state. Exiting the *pick task* via the *success* exit point results in a transition to the *place task*, in which the robot drives to the destination location and tries to place the object there. Major error conditions are handled explicitly via dedicated error states which result in retries of the same task or in cancellation of the current task and execution of a different one. Every other error, which can not be handled by error states, aborts the current task and leaves the state via the *failure* exit point. Transitions from all states to the *failure* exit point are omitted in the statechart in order to only show the main execution flow of the Pick And Place statechart.

During evaluation, the model shown in Table 5.2 was used in the simulation for predicting future resource utilizations. All numbers are estimates of the real resource usage of the actual components involved during the execution of the specific robot states.

Table 5.2: Resource models used during the simulation of the Pick And Place statechart.

State	**Memory [MByte]**	**CPU [%]**
Idle	50	1
Get Object Location	300	30
Move To Object Location	200	20
Move To Object Destination	200	20
Find Object	600	50
Visual Servo	300	80
Grasp Object	200	70
Lift Object	100	10
Find Place Position	300	50
Place Object	200	70

5.2.1 Single-Transition Prediction

Evaluation of the presented profiling architecture was performed by simulating the execution of the Pick And Place task with additional human intervention, as described in Section 5.2. Both spontaneous failure events as well as human interruptions were generated randomly throughout the execution of the whole task. Spontaneous failure events occurred with a probability of 0.3 during the execution of single sub-states.

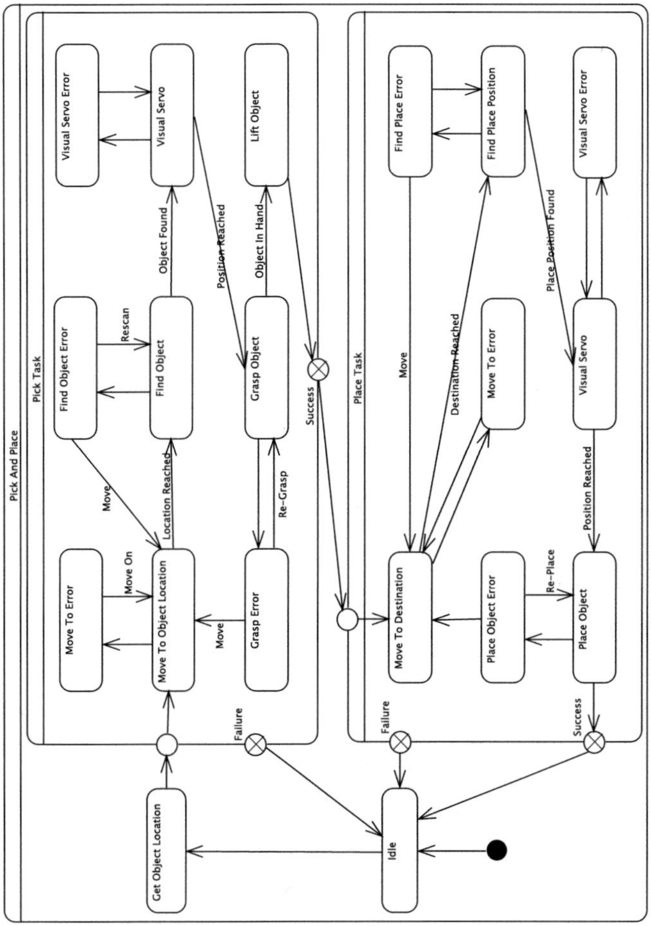

Figure 5.12: The *pick and place* statechart describing the tasks of the robot. The names of error handling states end with *error* and success is indicated by transitions to the *success* exit point. Transitions from any state to the *failure* exit point are hidden in the diagram to allow focusing on the main execution flow.

The human influenced the Pick And Place task according to the behavior and associated transition probabilities as shown in Figure 5.11.

In advance of the evaluation, four datasets for training the prediction model were gathered by repeated executions of the Pick And Place task. The datasets contain 50, 500, 1500, and 3000 recorded transitions respectively, including statechart and environment parameters. Each dataset with a lower number of transitions is a subset of the datasets with a higher number of transitions ($\forall i < j$: dataset i $\subset$ dataset j).

During the evaluation phase, the Pick And Place task was executed repeatedly until 500 transitions were reached. Each statechart transition instantly triggered a calculation of the state prediction for the next three future transitions using a prediction horizon of $h = 3$ time steps. After each state prediction, the accompanying resource utilization was calculated by the resource prediction and visualized.

Evaluation was performed using the Markov prediction model which is explained in Section 3.2.1. Additionally, the Global Probability (GP) method explained in Section 3.2.4 was used for comparison. The GP method takes only the global transition probability into account without regarding the world state. Prediction precision was determined by comparing the most probable predicted state (highest transition probability) with the actually executed state. Two different matching criteria were used during the evaluation. The first criterion measures the *State-Match* which only checks if the actually executed sub-state matches the predicted state. The second criterion measures the *World-Match* which compares both executed sub-state and environmental parameters. In the second case, predictions are only considered valid if both predicted and executed sub-state match as well as the environmental parameters Results of the evaluation are listed in Table 5.3 and shown in Figure 5.13.

Table 5.3: Prediction correctness in percent of the Markov and the GP prediction method. *State-Match* only detects sub-state matches while *World-Match* requires matches in both sub-state and environmental parameter. Values for GP and *World-Match* do not exist since this method does not take environmental parameters into account.

Dataset	Prediction Method	*State-Match* [%]	*World-Match* [%]
1	Markov	53.1	43.5
	GP	59.3	-
2	Markov	66.6	48.3
	GP	65.6	-
3	Markov	65.9	50.5
	GP	62.4	-
4	Markov	67.8	50.4
	GP	63.7	-

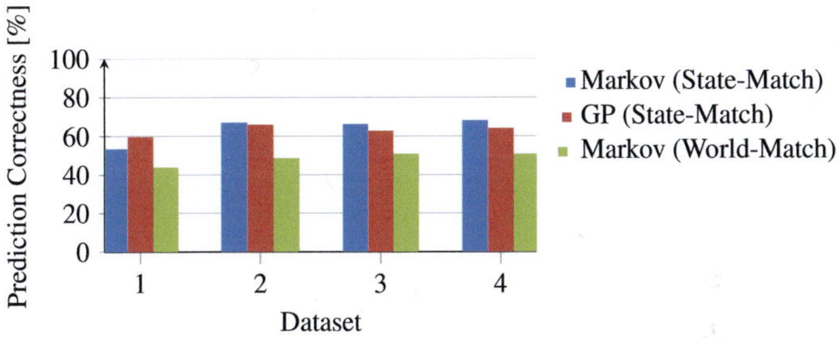

Figure 5.13: Prediction correctness percentage of the Markov and the Global Probability (GP) prediction methods as listed in Table 5.3.

For the smallest set of training data, the GP method shows a higher prediction rate than the Markov prediction. However, the Markov-based prediction performs better than GP after being trained with more data. With more training data, the Markov approach also has to rely less on the replacement strategy for unknown situations This trend is shown in Table 5.4.

Table 5.4: Percentage of required calls to the replacement strategy for the Markov prediction model based on the four different training datasets. With more training data, the model has to rely less on the replacement strategy, since more situations are stored in the Markov model.

	Dataset 1	Dataset 2	Dataset 3	Dataset 4
Replacement Stragety	58.9	42.6	29.3	25.0

The numbers in Table 5.3 show, that external interruptions lead to a failure rate of up to 47% (*State-Match* (SM)) or 57% (*World-Match (WM)*). Many interruptions can not be foreseen due to their random occurrence. Therefore, an evaluation was performed, where the Pick And Place task was executed without external interruptions while the model was trained with dataset 1. The comparison between Pick And Place execution with and without external events is listed in Table 5.5. The evaluations with failure events are marked with $+E$.

Table 5.5 shows that prediction performance is significantly higher, when no additional failure events are generated meaning the task is executed in a straightforward manner. One reason is that in the case of additional error events, at least one additional transition from the current state to the *failure* exit point is generated which is unknown in the trained model.

As expected, the *State-Match* criterion always produces more matches than *World-Match*. Prediction performance of the *World-Match* criterion decreases, since a specific robot task can be performed despite differing environment parameters. The robot might for example grasp the object whether a human is present or not. However, the *World-Match* measure is able to distinguish world states containing different humans, even if both humans perform the same task. It is also not possible to rely on *State-Match* in cases where interfering habits of different humans need to be considered. In the evaluation, it is essential to differentiate between the preferences of humans to grasp specific objects since the specific object determines if the robot will be interrupted or not. Increasing the number of correct predictions using the *World-State* criterion requires splitting environmental parameters into one class which influences the prediction result and another class containing all parameters which do not influence the prediction result.

Table 5.5: Percentage of correct predictions of the Markov based state prediction method. Evaluations marked with $+E$ contain failure events generated in addition to the human interruptions. *State-Match (SM)* only detects sub-state matches while *World-Match (WM)* requires matches in both sub-state and environment parameter. Due to the nature of the GP method, no *World-State* matches can be created.

Prediction Method	*SM* [%]	*WM* [%]	*SM* +*E* [%]	*WM* +*E* [%]
Markov	91.3	76.3	53.1	43.5
GP	79.6	-	59.3	-

5.2.2 Multi-Transition Prediction

In the previous section, prediction was performed for a single transition using a prediction horizon of $h = 1$ time steps. The same setup was used to evaluate the prediction performance for $h \in \{1, 2, 3\}$ time steps. Results based on the *State-Match* criterion are listed in Table 5.6. As expected, prediction accuracy declines for a prediction horizon of $h = 2$ and $h = 3$ time steps. Similar to results from the previous section, Markov produces better results than GP and the overall accuracy of the Markov prediction increases with the number of training samples used.

Table 5.6: Prediction accuracy for prediction horizon $h = 3$ time steps for Dataset 1 and Dataset 2 for prediction methods Markov and GP.

	Dataset 1			Dataset 2		
	h = 1	h = 2	h = 3	h = 1	h = 2	h = 3
Markov	53.1	35.2	28.9	66.6	49.6	36.7
GP	59.3	40.9	38.6	65.6	47.4	36,9

	Dataset 3			Dataset 4		
	h = 1	h = 2	h = 3	h = 1	h = 2	h = 3
Markov	65.9	47.7	37.3	67.8	50.8	43.7
GP	62.4	42.5	35.6	63.7	43.8	39.5

5.2.3 Resource Prediction

Predictions from the last section were further associated with the resource-profiles defined in Table 5.2. From each prediction calculated in Section 5.2.2, the most probable transitions with an accumulated probability greater or equal to 0.75 (75 %) were selected. Based on these transitions, three different measures of possible resource utilization are computed: lower boundary, upper boundary, and highest probability. Lower boundary describes the best case scenario, upper boundary the worst case scenario, and highest probability the scenario in which the most probable transitions are taken.

Three exemplary predictions and their associated resource predictions are shown in this section for the three states *visual servo*, *move to object location*, and *place object*. The statechart of the *pick task* is shown in Figure 5.14 and its sub-states *visual servo* and *move to object location* are highlighted in red.

The predicted transitions with the highest probability and their target state are highlighted in blue.

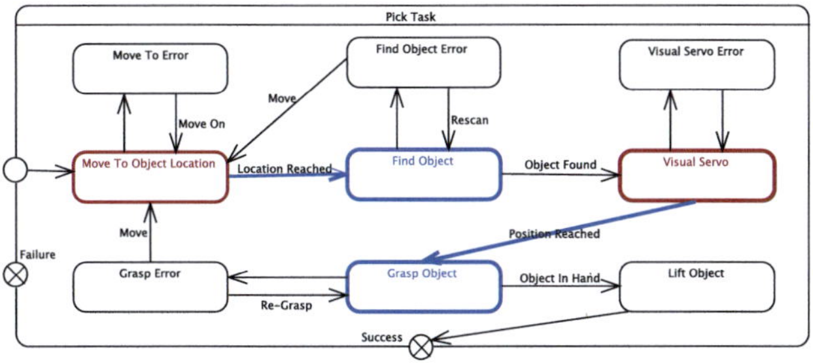

Figure 5.14: The *pick task* with its sub-states *pick object* and *move to location* highlighted in red. State transitions with the highest predicted probability are highlighted in blue.

The resource predictions for the states *pick object* and *move to location* are shown in Figure 5.15. CPU utilization is shown in the left graph while memory utilization is shown in the right graph of each sub-figure. Each graph shows the lower boundary of the predicted resource utilization as a red line, the upper boundary of the predicted resource utilization as a green line, and the predicted resource utilization with the highest probability is shown as a blue line.

The state *grasp object* was predicted to be the most probable successor state of the *visual servo* with an overall high CPU utilization (Figure 5.15(a)). The large difference between the upper and lower memory utilization boundary of step 1 is due to *failure* requiring less resources than *grasp error* which in turn requires less resources than the *lift object* state.

Both CPU and memory utilization of the resource prediction calculated from state *move to object location* start with an identical value of 50 % and 300 MByte, as shown in Figure 5.15(b). Based on the current world state, the only follow up state to be predicted from the current *move to object location* was *find object* state.

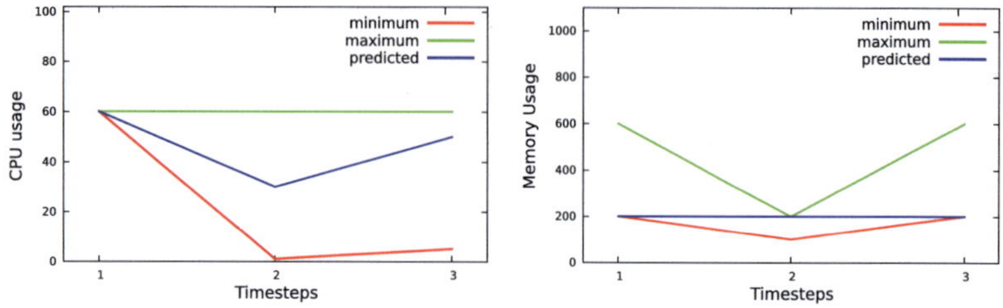

(a) Prediction for the *visual servo* state

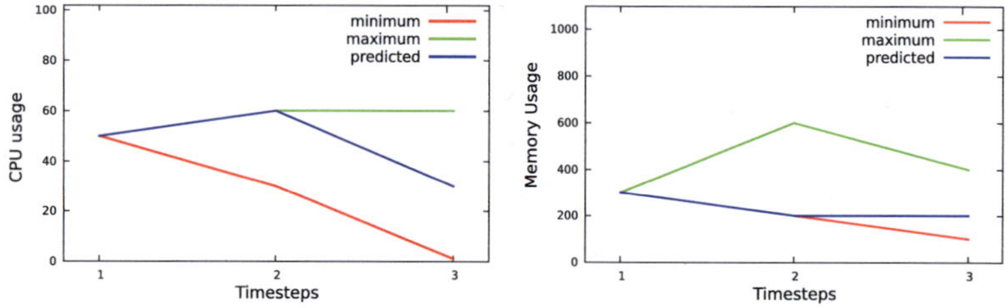

(b) Prediction for the *move to object location* state

Figure 5.15: Predictions for CPU and memory utilization generated for the *visual servo* and *move to object location* state. CPU utilization in % is shown in the left graph and memory utilization in MByte is shown in the right graphs of each sub-figure. Values colored in blue show the resource utilization of the predicted state with the highest probability. Furthermore, lower boundaries are shown in red while upper boundaries are shown in green. *Source:* [Kröhnert et al., 2014]

The *place task* is shown in Figure 5.16 and its sub-state *place object* is highlighted in red. The transitions with the highest predicted probability and their target state are highlighted in blue. The resource predictions for the state *place object* is shown in Figure 5.15.

Predictions for the *place object* show similar results to the *visual servo* state. In this case, the gap in CPU usage visible in Figure 5.17(a) results from the difference between *success* which requires no resources at all and the *place object* state which might be reentered after the occurrence of an interruption.

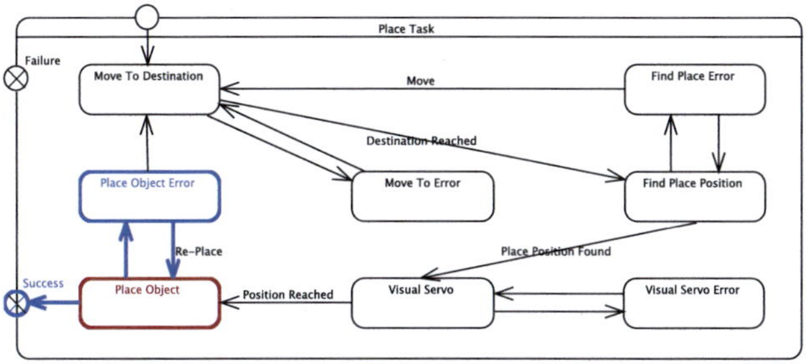

Figure 5.16: The *place task* with its sub-state *place object* highlighted in red. State transitions with the highest predicted probability are highlighted in blue.

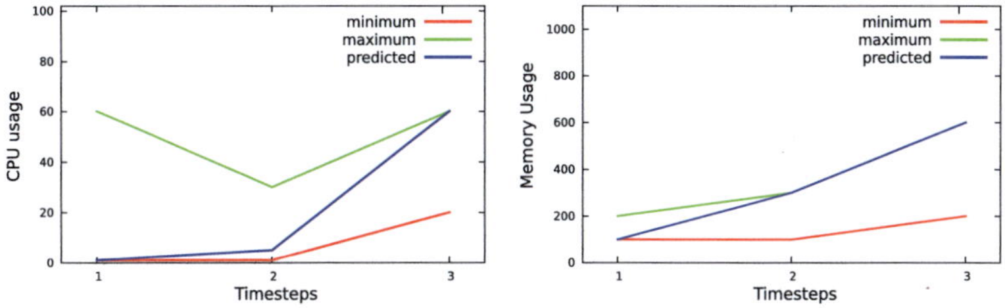

(a) Prediction for the *place object* state

Figure 5.17: Predictions for CPU and memory utilization generated for the *place object* state. CPU utilization in % is shown at in the left graph and memory utilization in MByte is shown in the right graph. Values colored in blue show the resource utilization of the predicted state with the highest probability. Furthermore, lower boundaries are shown in red while upper boundaries are shown in green. *Source:* [Kröhnert et al., 2014]

5.3 Resource-Aware Motion Planning

Evaluation of the two implemented resource allocation strategies is performed on four synthetic setups with increasing difficulty and one realistic setup with the humanoid robot ARMAR-4.

Each test case is executed 100 times and the algorithm runtime is measured in wall-clock-time. The bulk size parameter m is set to 10 to achieve the best results and was determined empirically. In these tests, no resources are removed from the algorithm at runtime to show the maximum performance of the algorithm. During planning of the test cases the manager process typically spends most of its time waiting for updates of its worker processes. Compared to the 100 % CPU utilization caused by the worker processes, the manager process only causes an insignificant amount of CPU utilization and is therefore excluded from the measurement. Finally, outliers are eliminated by using the 10 % trimmed mean of all values.

5.3.1 Test platform

The test cases are executed on up to four PCs connected via Gigabit Ethernet. An average CPU speed of 3.38 GHz is achieved with the following *Intel(R) Core(TM)* processors: 1 x *i7-4770* (3.4 GHz), 2 x *i7-4790* (3.6 GHz) and 1 x *i7 CPU 870* (2.93 GHz). All PCs were running the operating system *Ubuntu 12.04.5 LTS* which contains version 4.6.3 of the *g++* compiler. During test executions, the network was free from additional load and no other tasks were executed on the PCs alongside the motion planning algorithm. Since each *i7* processor contains 4 physical CPUs, each machine was allowed to start a maximum of four *RemoteObject*s (worker threads).

5.3.2 Test Case 1: SerialWalls

Performance of the resource-aware motion planning algorithm is evaluated with the *SerialWallsN* test cases which provide similar setups with increasing difficulty. This setup consists of $N \in \{1,2,3,4\}$ parallel walls with holes in the diagonally opposite corners, see Figure 5.18. In this setup, a long and thin object must be moved on a collision-free path from its start configuration ρ_{start} on the left side through the holes in the walls to the goal configuration ρ_{goal} on the right side of the walls. *SerialWalls4* represents the most difficult test case and is shown in Figure 5.18 together with the object to move and a computed collision-free path.

Overall system performance of the algorithm is evaluated using the following strategies:

- **Static Resource Allocation**: At the beginning of the test case, a constant number of worker processes is started subsequently. Solving the planning problem is immediately started by the first initialized worker. A fixed number of 1, 4, and 16 worker processes is used for evaluating the test cases.
- $\boldsymbol{\Delta T}$: The ΔT-strategy is evaluated for $\Delta T \in \{5, 10, 20, 30\}$. This strategy starts a new worker process after every ΔT seconds.
- $\boldsymbol{NN\Delta T}$: The $NN\Delta T$-strategy is evaluated with a fixed $\Delta T = 30$ and $\sigma \in \{1, 5, 10\}$ for a varying influence of the planning problem difficulty on the resource request frequency. The strategy dynamically adapts ΔT at runtime to request new worker processes when the planning progress is stagnating.

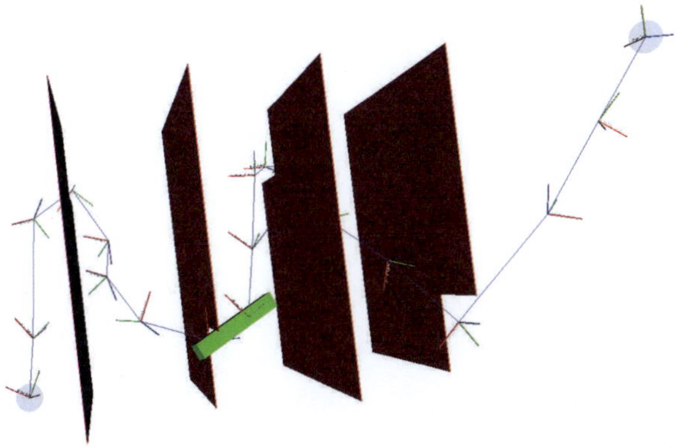

Figure 5.18: The test case *SerialWalls4* consists of four walls with opposing holes. The goal is to move the green beam from ρ_{start} on the left side of the walls to ρ_{goal} on the right side of the walls. *Source:* [Kröhnert et al., 2016] ©2016 IEEE

5.3.3 Test Case 2: Box

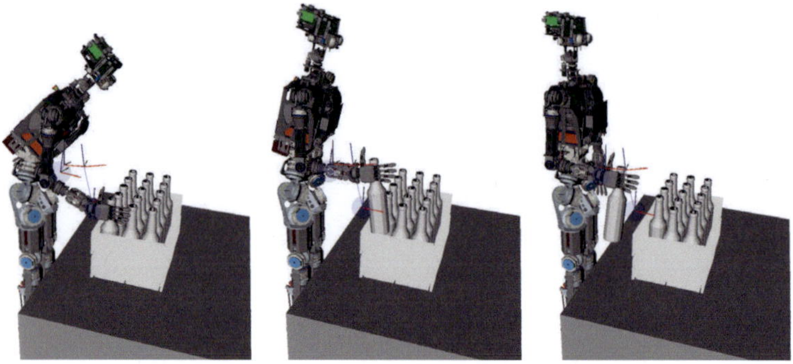

Figure 5.19: In the test case *Box*, the task of the humanoid robot ARMAR-4 is to pick up a bottle out of a box of bottles. The difficulty lies in getting the bottle out of the box without collisions, since it has to move in a straight line upwards. Collisions occur immediately if the bottle deviates from the straight path by a few of millimeters. *Source:* [Kröhnert et al., 2016] ©2016 IEEE

In the test case *Box*, a collision-free motion for the humanoid robot ARMAR-4 has to be planned for taking a bottle out of a box of bottles. Screenshots of this planning problem are shown in Figure 5.19. Both torso and the right arm are used for planning the motion, resulting in a total of 10 degrees of freedom (DoF). Two DoF are located in the torso while the right arm consists of 8 DoF. The motion contains a narrow and difficult passage at the beginning, where the bottle must be taken out of the box in a straight upward motion. Slight deviations in rotation and position of the bottle result immediately in collisions with the box.

This test case is used to examine whether the $NN\Delta T$-strategy can robustly compensate bad choices of the parameter ΔT. The ΔT-strategy is configured with $\Delta T \in \{5,30\}$, while $\Delta T = 30$ and $\sigma \in \{1,5,10\}$ were chosen as parameters for the $NN\Delta T$-strategy. Each test case initially starts with one worker process and is limited to request a maximum of 16 worker processes in total.

New worker processes are started first on any of the RemoteObjectNodes currently executing the least number of worker processes. Execution of the test cases is stopped after finding an initial solution, since the resource allocation strategies are only used during the initial planning phase.

Since the test case *Box* poses a difficult motion planning problem, the best strategy for finding a solution as fast as possible would be to start as many workers as fast as possible.

5.3.4 Static Resource Allocation Strategy

The static resource allocation strategy was tested on the *SerialWalls* test cases with the static number of workers set to $1, 4, 16$ and the number of walls $N \in 1, 2, 3, 4$. Figure 5.20 shows the result comparison of all combinations of N and number of static workers. The bars in the chart are normalized to the execution time t_{solve} of $N = 1$ static workers. Thus, the values for $N = 4$ and $N = 16$ show the speedup or how many percent faster the algorithm runs with more resources. Green bars show the normalized average execution time t_{solve}, which measures the time until the first solution was found. The time $t_{initial}$ measures the time required to start all initial workers and the resulting blue bars show the percentage of normalized $t_{initial}$ compared to normalized t_{solve}.

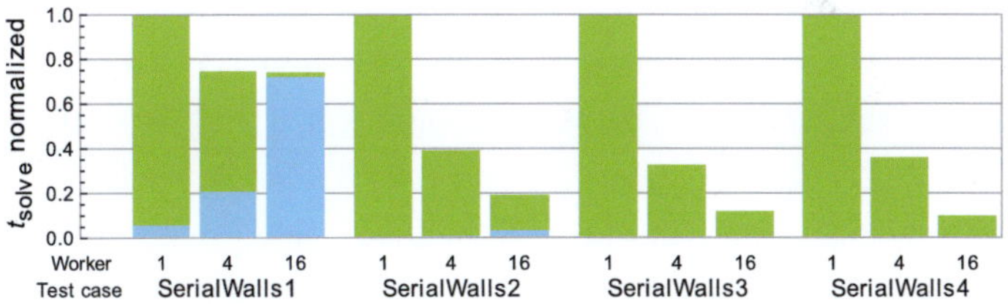

Figure 5.20: Normalized execution times for the *SerialWalls* test cases with static resource allocation. t_{solve} (green) represents the average execution time for finding the first solution and $t_{initial}$ (blue) shows the time required to start all initial workers. Detailed absolute planning times and their standard deviations are listed in Table 5.7. *Source:* [Kröhnert et al., 2016] ©2016 IEEE

Figure 5.20 shows, that increasing difficulty leads to a decrease in both t_{solve} and the ratio $t_{initial}/t_{solve}$ decrease if the number of static workers increases. However, a high number of static workers does not yield better results for simple problems as it can be seen in the *SerialWalls*1 test case. The graph shows that starting 16 worker processes requires almost as much time as four workers require for solving the *SerialWalls*1 planning problem. This is due to the static overhead of starting worker processes and copying data. During startup, all worker processes are started and initialized sequentially. Before a worker process starts the planning algorithm, the initialization routine must load all required collision models from the master process. Therefore, the time $t_{initial}$ is quite large for 16 worker processes to solve the simple *SerialWalls*1 test case. In this configuration, the last worker process has finished starting when the first worker has already found a solution and planning is stopped. The more difficult the motion planning problems get, the more negligible the static overhead gets, since t_{solve} increases with the difficulty but the time $t_{initial}$ stays mostly unchanged.

Absolute planning times and their standard deviation (*SD*) are listed in Table 5.7. It must be noted, that t_{solve} is measured in *seconds*, while $t_{initial}$ is measured in *millisconds* in order to not loose precision when displaying the standard deviation of both values with a limit of 3 numbers behind the decimal point. Since $SD_{t_{initial}}$ is smaller than $SD_{t_{solve}}$ by a factor of 1000, $SD_{t_{initial}}$ would only differ in the last number, thus not showing the variation in more detail. The tables contain an additional *Solved* metric, describing the ratio between succeeded vs overall planning attempts in a given time frame. For $N = 1$ the planning timeout was set to 12 minutes, while for $N = 4$ and $N = 16$ the timeout was set to 6 minutes. These timeout values were determined empirically and are shorter for more workers, since they solve the problem faster. With these timeouts, solutions were found in 100 % of all executions of the test cases *SerialWalls*1, *SerialWalls*2, and *SerialWalls*3.

However, test case *SerialWalls*4 could only be solved 42 % of the time with a single worker process, 94 % of the time with 4 worker processes, but 100 % of the time with 16 worker processes.

Table 5.7: The execution time average t_{solve} and standard deviation ($SD_{t_{solve}}$) until the first solution was found and the time $t_{initial}$ and standard deviation ($SD_{t_{initial}}$) required to start all initial workers.

Testcase	SerialWalls1			SerialWalls2		
Worker	1	4	16	1	4	16
Solved	100.0%	100.0%	100.0%	100.0%	100.0%	100.0%
t_{solve} $[s]$	0.950	0.701	0.700	21.100	8.200	4.000
$SD_{t_{solve}}$ $[s]$	0.006	0.031	0.009	1.103	0.397	0.131
$t_{initial}$ $[ms]$	56.600	200.700	687.500	59.900	218.100	768.900
$SD_{t_{initial}}$ $[ms]$	0.660	11.640	1.940	0.750	12.000	1.780
$t_{initial}/t_{solve}$	5.9%	28.3%	97.9%	0.3%	2.7%	19.2%

Testcase	SerialWalls3			SerialWalls4		
Worker	1	4	16	1	4	16
Solved	100.0%	100.0%	100.0%	42.0%	94.0%	100.0%
t_{solve} $[s]$	198.310	64.210	23.280	671.900	239.810	65.260
$SD_{t_{solve}}$ $[s]$	7.571	2.511	0.709	8.368	6.962	1.438
$t_{initial}$ $[ms]$	67.800	155.800	909.800	74.300	160.400	935.700
$SD_{t_{initial}}$ $[ms]$	0.740	6.390	1.850	0.880	1.150	1.420
$t_{initial}/t_{solve}$	0.0%	0.2%	3.9%	0.0%	0.1%	1.4%

5.3.5 Dynamic Resource Allocation Strategies

Both dynamic resource allocation strategies are first evaluated and compared in the context of the *SerialWalls* test cases. The following measures are used to obtain comparable numbers for all test cases:

- **Efficiency t_{total}**: The resource utilization efficiency can be expressed with t_{total}, the accumulation of each worker's individual execution time required to calculate the planning result (measured in each worker process). The smaller t_{total}, the more efficient the computation power is used. If only a single worker is started t_{total} equals the overall execution time.
- **Average Workload W_{avg}**: The average workload/number of workers (W_{avg}) is better comparable than the number of workers started in total throughout all executions (W_{max}). Since additional workers are requested at runtime based on a strategy, W_{max} proves to be an incomparable measure It does not account for varying amounts of required worker processes for processing the same problem multiple times.

$$W_{avg} := \frac{t_{total}}{t_{solve}} \tag{5.1}$$

t_{solve} describes the time required for finding the first solution (measured in the manager process). Thus, a value of $W_{avg} < 1$ is possible, since t_{solve} additionally includes the setup time of the master process which is for example spent to set up the planning environment.

Figure 5.21 shows a comparison of W_{max} and W_{avg} for all combinations of *SerialWalls* test cases and executed strategies. Detailed numbers are listed in Table 5.8. All dynamic resource allocation strategies started out with a single worker process and did not request any further resources for the *SerialWalls*1 test case. More resources were only requested by the dynamic resource allocation strategies for the more difficult test cases.

Figure 5.21 also shows that more worker processes are started with a lower value of the ΔT parameter. More worker processes are also started by the $NN\Delta T$-strategy, but with an increasing value of the σ parameter. For example, the $NN10\Delta T30$-strategy compensates for not optimally chosen high values of ΔT and starts almost double as many worker processes as the $\Delta T30$-strategy.

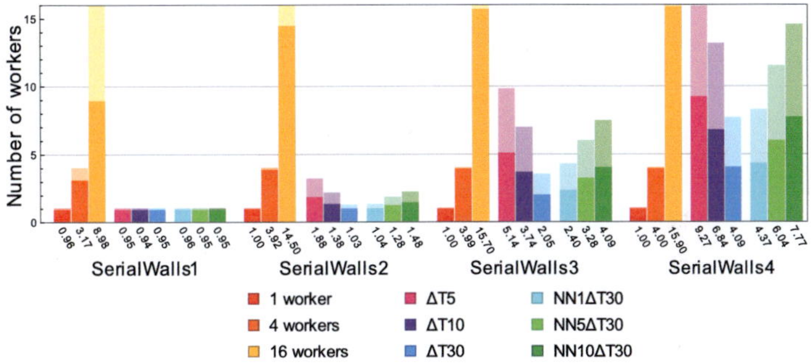

Figure 5.21: The number of workers allocated by the resource allocation strategies for each of the *SerialWalls* test cases. The bar of each strategy shows the number of total worker processes W_{max} in a half-transparent color and the average workload W_{avg} in a solid color. The value of W_{max} is always higher than the average W_{avg} if the two show different values. The figure shows that new resources are only allocated in case more difficult test cases such as *SerialWalls*3 and *SerialWalls*4 are encountered. *Source:* [Kröhnert et al., 2016] ©2016 IEEE

The effects of the resource allocation strategies on the execution time can observed by measuring and comparing t_{solve}. Normalized solving times are displayed in the left plot of Figure 5.22. The figure shows that *SerialWalls*2 is solved in the shortest amount of time with static resource allocation of 4 or 16 worker processes. However, the efficiency measure t_{total} is worse in comparison to the dynamic resource allocation strategies, since t_{solve} is partly spent on initializing worker processes.

Table 5.8: Number of total worker processes W_{max} and average workload W_{avg} for the graph shown in Figure 5.21. The first row lists the different strategies.

Testcase	SerialWalls1		SerialWalls2	
	W_{max}	W_{avg}	W_{max}	W_{avg}
1	1	0.96	1	1.00
4	4	3.17	4	3.92
16	16	8.97	16	14.51
$\Delta T5$	1	0.95	3.20	1.88
$\Delta T10$	1	0.94	2.16	1.37
$\Delta T30$	1	0.94	1.25	1.03
$NN1\Delta T30$	1	0.96	1.30	1.04
$NN5\Delta T30$	1	0.95	1.82	1.28
$NN10\Delta T30$	1	0.95	2.23	1.48

Testcase	SerialWalls3		SerialWalls4	
	W_{max}	W_{avg}	W_{max}	W_{avg}
1	1	1.00	1	1.00
4	4	3.99	4	4.00
16	16	15.72	16	15.90
$\Delta T5$	9.80	5.14	15.87	9.27
$\Delta T10$	6.96	3.74	13.11	6.84
$\Delta T30$	3.52	2.05	7.67	4.09
$NN1\Delta T30$	4.27	2.40	8.25	5.19
$NN5\Delta T30$	5.99	3.28	11.48	7.30
$NN10\Delta T30$	7.46	4.09	14.53	9.61

Normalized total execution times are displayed in the right plot of Figure 5.22. Particularly the setup with 16 static workers shows the worst efficiency for the *SerialWalls2* test case. In contrast, all dynamic resource allocation strategies run slower than strategies with static allocation but with a higher efficiency. Concrete numbers of the plots shown in Figure 5.22 are listed in Table 5.9.

Evaluation of the more difficult *SerialWalls4* problem shows that the dynamic resource allocation strategies perform almost as good as the static 16 worker strategy while the measured efficiency of the dynamic strategies remains better.

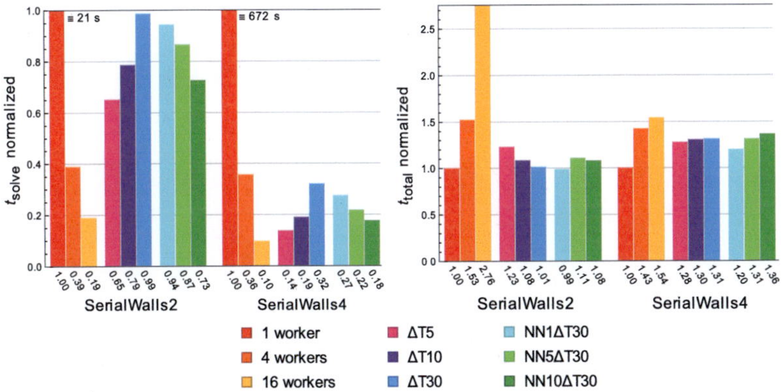

Figure 5.22: Execution time t_{solve} and efficiency t_{total} for the test cases *SerialWalls2* and *SerialWalls4* per evaluated resource allocation strategy. All measured values are normalized to the execution time of a single worker. See Table 5.9 for concrete numbers. *Source:* [Kröhnert et al., 2016] ©2016 IEEE

In general, the efficiency of all dynamic resource allocation strategies is comparably better than the efficiency of static approaches with more than 1 worker process. Therefore, the implemented dynamic resource allocation strategies utilize the available computation power more efficiently, especially when applied to easier planning problems. The difference in efficiency between static and dynamic resource allocation strategies gets smaller with increasingly more difficult motion planning problems.

Table 5.9: Corresponding Execution time t_{solve} and efficiency t_{total} numbers for the test cases *SerialWalls*2 and *SerialWalls*4 per evaluated strategy as shown in Figure 5.22. The first row lists the different strategies.

Testcase	SerialWalls2		SerialWalls4	
	t_{solve} [s]	t_{total} [s]	t_{solve} [s]	t_{total} [s]
1	21.1	21.1	671.9	672.0
4	8.2	32.1	239.9	959.0
16	4.0	58.0	65.3	1,040.0
$\Delta T5$	13.7	25.9	92.5	857.0
$\Delta T10$	16.6	22.8	127.9	875.0
$\Delta T30$	20.8	21.3	215.6	882.0
$NN1\Delta T30$	19.9	20.8	184.4	805.0
$NN5\Delta T30$	18.3	23.3	146.0	881.0
$NN10\Delta T30$	15.3	22.7	118.0	917.0

5.3.6 Evaluation on the Humanoid Robot ARMAR-4

A realistic motion planning problem is posed by the *Box* setup in which the task of the humanoid robot ARMAR-4 is to take a bottle out of a box of bottles (Figure 5.19). This test case was selected to show if and how well a bad choice of the parameter ΔT can be compensated by different choices of the parameter σ of the $NN\Delta T$-strategy.

The metrics W_{avg}, W_{max}, t_{solve}, and t_{total} are shown in Figure 5.23 and the concrete values are listed in Table 5.10. For all test runs the execution time was limited to a total of 6 minutes. The execution of the static configuration with one worker process is not shown in the diagrams, since it only found a valid path within the given time limit for 33% of the test runs.

Figure 5.23 shows that the $\Delta T30$ strategy obviously starts the least amount of workers compared to all other strategies, due to the bad choice of $\Delta T = 30$. The runtime of $\Delta T30$ is more than three times larger than the runtime of $\Delta T5$ and the efficiency t_{total} is two times worse.

All $NN\Delta T$-strategies compensate the high ΔT value and start more workers than the $\Delta T30$ strategy, resulting in lower t_{solve} and t_{total} values. Compensation of high values of ΔT is stronger with an increasing value of σ.

$NN10\Delta T30$ solves the problem 15% faster than $\Delta T5$ but with a 7% lower efficiency value t_{total}. These results again show the ability of the dynamic resource allocation strategies to detect difficult planning problems. Additionally, the requested computation resources are utilized efficiently and the $NN\Delta T$-strategy shows its ability to compensate bad choices of the ΔT parameter.

Table 5.10: Average number of workers W_{avg} and maximum number of workers W_{max} for the test case *Box* and all employed resource allocation strategies. Average execution time t_{solve} and the associated standard deviation $SD_{t_{solve}}$ for the test case *Box* and all employed resource allocation strategies. Total execution time t_{solve} and the associated standard deviation $SD_{t_{solve}}$ for the test case *Box* and all employed resource allocation strategies.

	$\Delta T5$	$\Delta T30$	$NN1\Delta T30$	$NN5\Delta T30$	$NN10\Delta T30$
W_{avg}	5.7	3.5	4.3	5.6	7.0
W_{max}	10.1	6.7	8.1	10.4	11.9
t_{solve} [s]	58.2	187.8	127.2	75.0	49.7
$SD_{t_{solve}}$ [s]	3.8	7.3	5.2	3.7	3.0
t_{total} [s]	324.0	656.0	544.0	420.0	348.0x

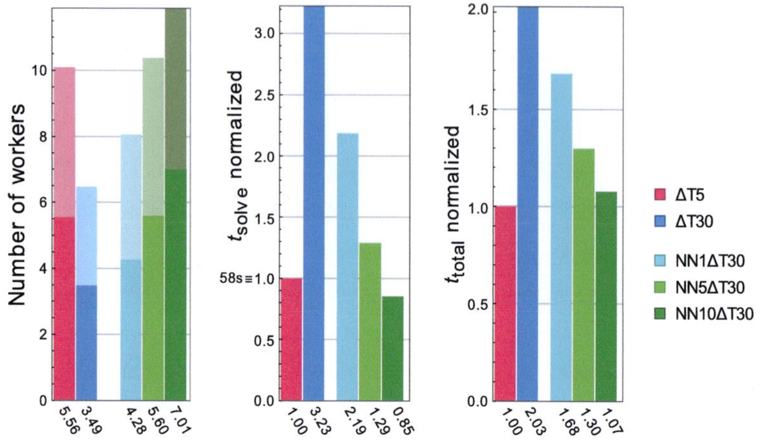

(a) W_{avg} and W_{max} (b) t_{solve} (c) t_{total}

Figure 5.23: W_{avg}, W_{max}, normalized t_{solve} and normalized t_{total} for the test case *Box* per evaluated strategy. *Source:* [Kröhnert et al., 2016] ©2016 IEEE

5.4 Resource-Aware Disparity Map

The resource-aware disparity map algorithm is evaluated as X10 implementation using the invasive simulator and as C++ implementation running on top of OctoPOS on an Invasive Computing hardware prototype. Furthermore, the integration of the disparity map algorithm into a demo setup is shown.

5.4.1 Invasive X10 Evaluation

Evaluation of the algorithm was performed using the invasive simulator [Hannig et al., 2011]. The simulator was executed on a PC equipped with a Core i7 (4 cores with HyperThreading @ 2.93 GHz) processor running Ubuntu Linux 8.04 and X10 version 2.2.1. Both the X10-runtime using the C++ backend as well as the invasive simulator are compiled from scratch with the compiler flags `-Doptimize=true -DNO_CHECKS=true` to achieve the best performance.

Overall, execution time is measured in 5 trials, each running the algorithm 100 times using the same set of input images. The resource-aware *disparity map* algorithm is executed using 1 to 8 PEs, while the IVT and the single-core X10 version are not parallelized and can therefore not use more than one PE.

Evaluation results are shown in Table 5.11 which lists the IVT and single-core X10 versions first, followed by the resource-aware version. The fastest evaluated variant was the IVT implementation. Slowdowns in all other variants are due to the used X10-runtime not containing fast and efficient implementations for accessing *Array* data structures.

As expected, the invasive variant using one PE shows slightly slower performance than the plain X10 and the C++ version. In particular, the other variants do not use the `invade()`, `infect()`, `retreat()` API for acquiring and releasing resources and therefore do not require copying data over the NoC into remote TLMs. Hence, the invasive API and simulator overhead is comparatively small and accounts only for a small overhead in execution time.

Overall, it can be shown that the performance of the invasive disparity map algorithm measured in execution time increases with the number of PEs available.

Table 5.11 also shows the range of calculated standard deviations for all test executions. The usage of Intel Core i7 processors with HyperThreading technology can most probably be made responsible for these ranges. HyperThreading allows simultaneously scheduling two operating system threads on the same CPU core. This technique can boost overall system performance, but can also reduce performance if simultaneously executing threads share the same set of processing resources. In the second case, one thread has to wait until the shared resource is freed again by the other thread.

Table 5.11: Performance evaluation of the different disparity map implementations showing the number of used PEs and the mean, min, max, and standard-deviation (SD) of measured execution times.

Algorithm	#PEs	Mean [ms]	Min [ms]	Max [ms]	SD [ms]
IVT (C++)	1	38	37	90	3-9
X10	1	93	84	143	11-17
Invasive	1	135	125	236	3-14
Invasive	2	138	125	193	3-11
Invasive	3	143	104	283	6-18
Invasive	4	144	131	283	3-19
Invasive	5	120	89	150	4-6
Invasive	6	105	98	134	3-6
Invasive	7	96	65	148	4-8
Invasive	8	90	81	213	3-18

5.4.2 Invasive C++ Evaluation

Evaluation of the OctoPOS native implementation of the disparity map was performed on the hardware shown in Figure 4.5. Stereo images of the size 320x240 pixels were used as input for producing a same-sized disparity map. The invasive C++ implementation of the algorithm was executed in two different configurations both using between 1 to 6 CPUs. First, the algorithm uses main memory for storing input, intermediate, and final data structures. Second, tile-local memory (TLM) is used for storing intermediate data structures used for local processing.

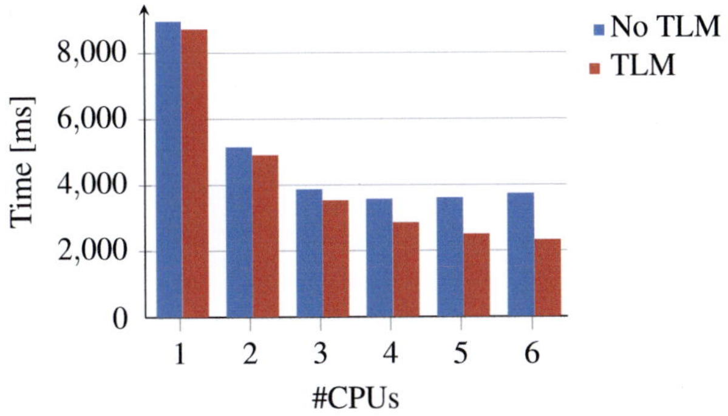

Figure 5.24: Execution times in [ms] of the invasive C++ disparity map algorithm. Blue bars represent executions without using tile-local memory (TLM). Red bars represent executions making use of TLM.

Figure 5.24 shows a graph of the results with detailed execution times listed in Table 5.12. The graph shows that on physical hardware, execution times decrease with a higher number of used CPUs. It is also shown that storing intermediate data in available TLM always results in faster execution times as compared to storing this data in the main memory. This reduction in execution time is due to the faster access to the TLM and a reduction of the bus load of the bandwidth limited connection between CPUs and main memory.

Table 5.12: Execution times in [ms] of the invasive C++ disparity map algorithm. The top row indicates the number of CPUs used for calculation. Execution times in the first data row are without using tile-local memory (TLM). Results in the second data row were obtained using TLM.

	1	2	3	4	5	6
No TLM [ms]	8950	5130	3850	3540	3580	3700
TLM [ms]	8700	4890	3500	2830	2480	2300

5.4.3 Invasive Computing Head Demo

Exemplary screenshots of the demo execution are provided in Figure 5.25 and Figure 5.26. The screen is divided into four different images showing: one of the input images (top left), the calculated disparity map image (bottom left), object detection (top right), and skin detection (bottom right). The first image (Figure 5.25(a)) shows an empty scene with uninteresting objects in the background, resulting in computations of the disparity map. Skin color is first detected in a low-resolution version of the input images as seen in Figure 5.25(b). Calculation of the disparity map continues in this case. Once, the target object is detected in a low-resolution image, the detection algorithm requests more resources to track both the object as well as any skin colored region with higher quality, see Figure 5.26(a). In this case, not enough resources remain for calculating the disparity map image.

(a) No objects in scene.

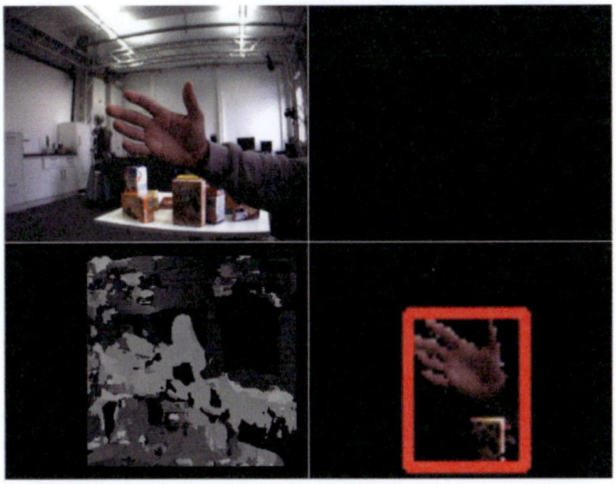

(b) Disparity map plus low-resolution skin detection.

Figure 5.25: Invasive tracking demo screenshots, showing an input image (top left), disparity map image (bottom left), object tracking output image (top right), and the skin tracking output image (bottom right). Only the disparity map is calculated if no objects or skin is detected (5.25(a)). Object recognition performed on low-resolution (5.25(b)).

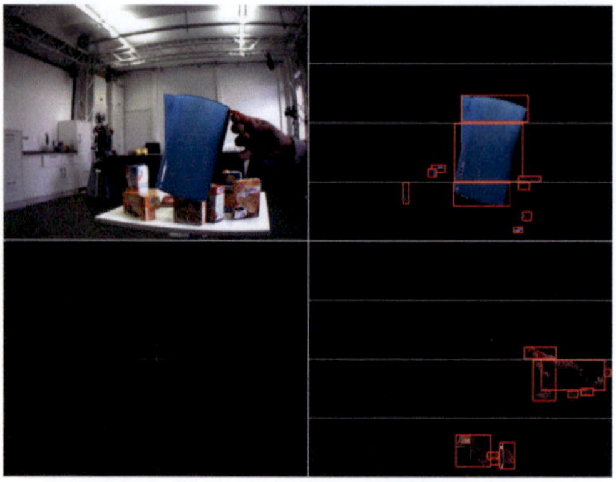

(a) High-resolution object and skin detection.

Figure 5.26: Invasive tracking demo screenshots, showing an input image (top left), disparity map image (bottom left), object tracking output image (top right), and the skin tracking output image (bottom right). Object recognition is performed on full-resolution images (5.26(a)) after an object was recognized in a low-resolution image (5.25(b)).

6 Conclusion

The objective of this work was to provide building blocks for future resource-aware architectures for humanoid robots. Since building a complete resource-aware architecture is a huge task in itself, the scope of the work was to predict resource over-utilization at runtime and to assign resources to concurrent applications in a context-sensitive manner. This led to the key contributions of this work, namely, the creation of context-sensitive resource models, a prediction mechanism for future robot tasks, and the prediction of future resource utilizations when resource models are combined with robot task predictions. Resource-aware algorithms were created, that are capable of adjusting their internal resource usage based on the amount of available and assigned resources.

In summary, the contributions of this work to a resource-aware robot architecture for humanoid robots are the following:

Context-Sensitive Resource Models

Humanoid robots work in dynamically changing environments where executions of tasks can differ significantly based on the current context. This thesis presented a context-sensitive resource model which can be generated from data extracted from real or simulated executions of robot tasks which are described as statecharts. The model can specify varying resource demands and execution times based on environmental contexts.

Furthermore, a profiling architecture which allows runtime introspection and self-monitoring of robot applications and statecharts was introduced and integrated into the existing ArmarX robot framework.

This architecture enables collecting high-level information about control and data flow within robot application statecharts as well as the environment of the robot. Runtime profiling of robot applications makes this information available as training data for the resource models. The acquired data is also stored in a database for further offline processing.

Context-Sensitive Resource Prediction

An architecture for predicting future robot tasks and resource utilizations based on the current context was introduced in chapter 3. The components required for realizing the prediction architecture were implemented as enhancements of the existing ArmarX robotics framework and the previously described profiling mechanisms were reused for runtime monitoring purposes. A prediction model based on Markov chains was trained with data from simulated executions of ARMAR-III performing a *pick and place* task in a fully observable environment. During the task, the robot was randomly disturbed by human interactions and external failure events. The model was shown to be capable of calculating probability distributions of the most probable next robot tasks depending on the current world state. Predictions of future robot tasks were further enriched with information from the presented context-sensitive resource models to provide probabilistic information about upcoming workload situations and their expected resource utilization.

The evaluation showed that the profiling architecture can also be used to monitor environment and transition parameters of ArmarX-based robot programs. During execution, predictions were constantly calculated upon changes in the robot statechart and enhanced with context-sensitive resource models. The presented approach was able to provide the means for detecting resource over-utilizations in advance of their occurrence. This resource prediction information provides the basis for performing speculative resource management, either through extensions of off-the-shelf operating systems or in combination with specialized resource-aware operating systems.

Currently, it is difficult to estimate the benefit of using prediction information in operating systems. In general, prediction is not used in operating systems, except in research-oriented operating systems. Additionally, dynamic resource profiles of applications are usually not taken into account by resource allocation mechanisms of existing operating systems. Most current systems lack the possibility to express these dynamic resource demands of applications and to take these demands into account during resource distribution.

One goal of new paradigms such as Invasive Computing is to prevent system overload situations before they occur by dynamically influencing the resource negotiation process in advance of actual resource requests. However, taking advantage of these resource negotiation capabilities requires algorithms to be capable of dynamically adjusting their internal resource demands based on resource availability.

Resource-Aware Motion Planning

State of the art humanoid robots are equipped with multi-core processors, which are used either to execute parallelized algorithms or for parallel execution of different tasks and computations. Motion planning is one category of algorithms benefiting from parallelization. However, current parallelized and distributed motion planners work with a statically assigned number of resources in order to provide results for difficult planning problems as fast as possible.

This work showed that static resource allocation for motion planning algorithms is not optimal for solving planning problems of various difficulties. Additionally, static allocation can lead to inefficient resource utilization and result in an increased system workload. To overcome this problem, a distributed resource-aware motion planning algorithm was presented which was implemented in the ArmarX robotic framework.

This motion planning algorithm starts with minimal resources and is capable of requesting additional computing resources based on the progress and the difficulty of the current motion planning problem. Different dynamic resource allocation strategies for tracking the planning progress were implemented. The strategies were evaluated in increasingly difficult synthetic motion planning setups and in a realistic motion planning scenario involving the humanoid robot ARMAR-4. These resource allocation strategies were shown to provide a good estimate of the planning problem difficulty which could be translated into a measure for requesting additional resources. Resource usage efficiency was increased by acquiring resources only for difficult problems. However, the increased resource utilization comes at the cost of a slightly increased overall execution time. Therefore, the planner parameterization poses a trade-off between efficiency and speed.

Resource-Aware Disparity Map

In the field of robotic vision, a resource-aware disparity map algorithm based on the new Invasive Computing paradigm was developed in this work. The initial version was implemented in X10 and evaluated using a simulator which provides support for invasive techniques for dynamically requesting and releasing resources of a small CPU cluster. This version of the algorithm was shown to be capable to dynamically adapt to different amounts of available resources. Afterwards, the algorithm was implemented in C++ on top of OctoPOS, the operating system of the Transregional Collaborative Research Center Invasive Computing (SFB/TR 89), and evaluated on prototype hardware. It could be shown that utilization of features of the Invasive Computing hardware platform results in further speedup of the algorithm. Such algorithms, which can dynamically scale their performance and functionality depending on the current state of the robot, are intrinsic parts of future resource-aware robot architectures which execute on resource restricted platforms.

6.1 Outlook

This thesis focuses on basic building blocks required for building resource-aware robot architectures. Namely, a context-sensitive resource model, a profiling approach for recording execution data, an approach for predicting future robot tasks and associated resource utilizations, and resource-aware algorithms. These components provide the basics for performing speculative resource management on humanoid robots. Future work therefore includes the expansion of the presented *pick and place* scenario to use predicted resource utilizations in order to speculatively start computations before they are requested. Furthermore, both current and predicted resource utilization data can be used to influence resource-aware algorithms at runtime by granting more resources if available or by withdrawing resources when other applications request resources. For example, motion planning could be speculatively started earlier than requested using as many resources as available at any time. This scenario would also serve as an evaluation of the resource-aware motion planner to be capable of handing back resources to balance the system workload while still being capable of fulfilling the planning request.

Currently, the prediction approach requires full knowledge of the environment, and thus a fully observable world. This is however not ideal for humanoid robots operating in changing environments. A future enhancement would be to address this issue by evaluating prediction models such as HMMs, which are capable of dealing with incomplete world knowledge.

The context-sensitive resource model includes both CPU and memory utilization statistics as well as statistics of execution time and branching behavior. A future task related to this model is to include communication bandwidth and behavior which also requires enhancements to the profiling infrastructure.

One further aspect is automatic detection of ”relevant“ components or states in a statechart which should be included in the profiling process. By not processing every piece of information, the amount of gathered and stored data could be reduced.

The resource-aware motion planner can be expanded with more specialized resource allocation strategies.When requesting more resources, these strategies might for example take the current system load, memory pressure, or communication bandwidth into account. Also, investigating and decreasing the startup time of newly created worker nodes can result in reducing the gap in execution times when many workers are used.

Appendix

A Markov Processes and Markov Chains

Markov processes describe statistical models that use probabilities to describe possible occurrences of future events. One use case is stochastic modeling of environmental world states. Two different kinds of Markov processes exist

- discrete-time: transitions within the process happen at discrete time steps, meaning that time steps are countable
- continuous-time: time steps are not countable

Discrete-time Markov processes are also called Markov chains and are described in more detail in this section. One major aspect of this type of model is that limited knowledge of the past is sufficient to calculate the probabilities for the next transitions in the model. The Markov property states that the probability distribution of future states in the process only depends on the current state of the process. This Markov property holds true for first order Markov chains, where only the current state of the model influences future transitions. In contrast, the probability distribution of nth order Markov chains is influenced by the past n states occurred in the process. Conditional probabilities are used to model this relation mathematically.

A Markov chain is based on a finite set of states $S = \{s_1, \dots, s_m\}, m \in \mathbb{N}$. Each state can represent a robot pose, a robot task, or emotional states of a human. Transitions between states in the Markov chain are triggered by observable events. Markov chains are described by a sequence of random variables $X_1, \dots, X_m, m \in \mathbb{N}_0$ with the property

$$P(X_n = s_n | X_0 = s_0, X_1 = s_1, \dots, X_{n-1} = s_{n-1}).$$

Here, $P(X_n = s_x | X_{n-1} = s_y)$ describes the probability of the transition $(s_x \rightarrow s_y$ at time step n. Using $p_{ij}(n) = P(X_n = j | X_{n-1} = i)$ as abbreviations leads to the following compact matrix representation of the transition probabilities

$$\Pi_n = \begin{pmatrix} p_{11}(n) & \cdots & p_{1m}(n) \\ \vdots & \ddots & \vdots \\ p_{m1}(n) & \cdots & p_{mm}(n) \end{pmatrix}.$$

A Markov chain is called homogeneous, if every transition probability is independent of the time parameter n resulting in the following property

$$P(X_n = s_x | X_{n-1} = s_y) = P(X_m = s_x | X_{m-1} = s_y).$$

The advantage of a homogeneous Markov chain is the possibility to calculate the transition probability from $(s_x \rightarrow s_y)$ via $p^n_{ij} = P(X_{k+n} = s_y | X_k = s_x)$.

Initially, a state distribution $p(0)$ is required, describing how probable it is for each $s \in S$ to occur at time step 0. The product of $p(0)$ and the transition matrix Π then provides the probability distribution of all states in the process after one time step.

Furthermore, Markov chains can be visualized intuitively as directed graphs. An exemplary graph is given on the following page. The matching transition probabilities are listed in the table after the figure.

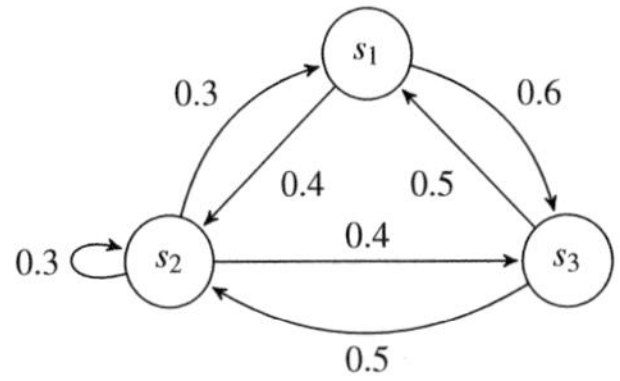

A Markov model consisting of three states s_1, s_2, s_3, transitions between the states, and associated transition probabilities.

Transition probabilities of the Markov model depicted in the figure above.

	s_1	s_2	s_3
s_1	0	0.4	0.6
s_2	0.3	0.3	0.4
s_3	0.5	0.5	0

B Voronoi Diagram and Voronoi Regions

A Voronoi diagram divides a plane into partitions based on seed points. Given a set of seed points $p_1, p_2, \cdots, p_n$ the partition R_{p_i} associated with seed point p_i is constructed as follows. The distance between each point p and the seed point p_i is smaller than the distance between p and any other seed-point p_j $(i \neq j)$. The partition R_{p_i} is also called the Voronoi region of seed point p_i.

Different examples of Voronoi diagrams are shown in the figures on the next page. The left image shows a square plane divided vertically into two equally sized partitions. The middle image shows the same plane divided into four equally sized partitions through four seed points. The locations of the seed points in the left and middle image lead to the regular geometric shape of the Voronoi regions. An example of a more complex and geometric irregular Voronoi diagram is shown in the right image, which divides the plane into 14 differently sized partitions.

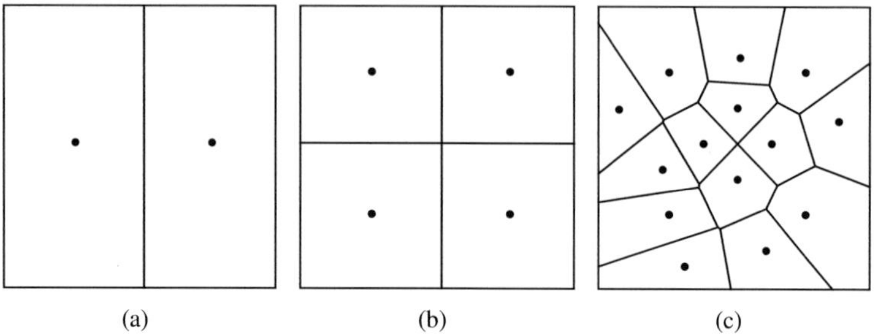

Voronoi diagrams for 2, 4, and 14 seed points.

C ArmarX

ArmarX is an event-driven Robot Development Environment (RDE) providing the infrastructure for creating distributed component-based software architectures.

Communication in ArmarX is performed by the Internet Communication Engine (Ice) framework provided by the company ZeroC [ZeroC, Inc., 2015]. The foundations of Ice include an Interface Definition Language (IDL) and different communication patterns. IDLs allow specifying interfaces in a type-safe manner and creating software using different programming languages by transforming the IDL constructs into each language. Communication patterns provided by Ice are remote procedure calls and publish-subscribe. Remote procedure calls are used for direct interaction with specific software components and can be one-way for commands or two-way if a result is expected. Publish-Subscribe is used when data such as sensor values need to be send to an unspecified number of receivers. In this case, the data broadcast by the publisher on so called topics is received by a server which in turn delivers the data to receivers which have subscribed to specific topics. Additionally, all communication using Ice can be changed between asynchronous or non-blocking based on context dependent requirements.

Organization of ArmarX can be divided into three major categories. The middleware category provides core functionalities such as communication, deployment, the component model, dependency management, and the statechart application model. The robot framework category provides standardized interfaces and components which can be reused and configured for a specific robot. Components of the MemoryX memory architecture or components for working with the kinematics of a robot are examples of such reusable and reconfigurable components. The application category contains all software implemented for a specific robot. Extensive introspection, debugging, and development capabilities are provided by supporting tools such as generic and specialized visualizations and editors or the ArmarXSimulation.

C.1 ArmarX Statecharts

The application model of ArmarX consist of two parts: robot components and statecharts. Robot components perform most of the work such as accessing hardware, computing kinematics, or detecting objects. Statecharts on the other hand represent the high-level structure of robot programs and specify control flow, data flow, and the coordination logic of robot components.

The major design goals of the ArmarX statecharts are modularity, reusability, runtime-reconfigurability, decentralization, and state-disclosure. Modularity and reusability are tightly coupled, meaning that states themselves are decomposed into smaller statecharts with each statechart being reusable in other statecharts. Runtime reconfigurability provides the means to dynamically change the structure of a statechart, an essential feature for executing dynamically changing plans. Decentralization of statecharts allows executing sub-statecharts in a process different than the parent process, leading to more robust systems capable of load balancing and crash recovery from faulty lower-level statecharts. In addition to that, state-disclosure allows inspecting the structure, the current state, and parameters of statecharts at runtime. This feature aids developers in debugging control- and data flow issues at runtime.

A state in an ArmarX statechart is defined by a set of input parameters, a set of local parameters, a set of output parameters, a set of outgoing events, as well as a collection of sub-states and parameter mappings between sub-states. Additionally, user specified code can be executed when entering or leaving a state. Different types of states exist: regular local states, end states for triggering an event and leave the parent state, remote states for accessing states running in a remote process, and dynamic remote states which are similar to remote states but instantiate the target state at runtime.

Transitions are defined between a source and a destination state on the same hierarchy level and are triggered by events generated by end states or by conditions. Conditions are logical expressions defining checks on input data and an event to fire when evaluating to true. In addition to the control flow, data flow is specified via transitions by mapping source state output parameters and parent state parameters to destination state input parameters. A special initial state triggers a transition to a predefined sub-state when a statechart is entered.

To enhance reusability of statecharts between different use cases, so called statechart profiles are available. These statechart profiles allow changing default values of state parameters based on the chosen profile. This feature provides the means for creating reusable and more generic statecharts, since robot dependent parameters can be optimized through changes in the robot specific statechart profile.

Development of ArmarX statecharts is aided by the StateChartEditor (SCE) which allows creating and connecting statecharts graphically via a drag and drop interface. The SCE allows defining typed input, output, and local parameters, specifying and overriding statechart profiles as well as mapping state parameters based on transitions. Furthermore, the SCE contains sanity checks to enforce the user to only provide parameter mappings between compatible types.

C.2 MemoryX

MemoryX provides a memory architecture for robots inspired by the human brain. The architecture consists of three different types of memory: Working Memory (WM), Long-Term Memory (LTM), and Prior Knowledge (PK). All memory types are accessible via Ice throughout the complete application network which allows querying and updating the memory from all parts of the distributed robot application. Each memory type is divided into segments for grouping semantically similar data such as perceived objects, known object classes, or locations of robots and other agents in the world. Information is stored in these memory types as entities representing generic key-value pairs. Specialized entity classes exist holding a predefined set of keys. Further entity classes can be specified by the robot programmer.

The Working Memory is a volatile memory, similar to the human's short-term memory, and contains the current robot's internal state and all information about the current environment known to the robot. Updates to the Working Memory are usually performed by perception processes which instantiate object classes found in Prior Knowledge.

Prior Knowledge contains persistent information provided by the user or designer of the robot. Entities stored in Prior Knowledge may contain 3D object models or visual features required for object detection. Furthermore, entities in Working Memory can be enriched with information from Prior Knowledge for example to display 3D objects in a visualization.

The Long-Term Memory is used to persistently store information learned or inferred by the robot. Furthermore, snapshots of the Working Memory can be created and stored in LTM. These snapshots can be used to store predefined environments which can be loaded by the robot into Working Memory.

This memory architecture combined with perceptual components provide a flexible and easy to use framework. Instead of directly querying the perception about locations of objects, the memory architecture is asked for this information. If the existence certainty of an object is high enough, the information is directly returned. If the existence certainty is below a specified threshold, the respective memory triggers the associated object detection components and fuses the returned result with the existing entities. MemoryX provides different motion models which can be attached to entities. These motion models are used to predict where a moving object will most likely be found in the future. Additionally, MemoryX allows registering for events which are sent out if specific objects or the general content of the memory changes.

List of Figures

List of Tables

List of Algorithms

Bibliography

O. Adiyatov and H. A. Varol. Rapidly-exploring random tree based memory efficient motion planning. In *2013 IEEE International Conference on Mechatronics and Automation (ICMA)*, pages 354–359, Aug. 2013.

R. Albers, E. Suijs, and P. de With. Resource Usage Prediction for Groups of Dynamic Image-Processing Tasks Using Markov Modeling. In *2009 IEEE International Conference on Acoustics, Speech and Signal Processing. ICASSP 2009*, pages 1929–1932, Apr. 2009.

T. Asfour, K. Regenstein, P. Azad, J. Schroder, A. Bierbaum, N. Vahrenkamp, and R. Dillmann. ARMAR-III: An Integrated Humanoid Platform for Sensory-Motor Control. In *2006 6th IEEE-RAS International Conference on Humanoid Robots*, pages 169–175, Dec. 2006.

T. Asfour, K. Welke, P. Azad, A. Ude, and R. Dillmann. The Karlsruhe Humanoid Head. In *Humanoids 2008 - 8th IEEE-RAS International Conference on Humanoid Robots*, pages 447–453, Dec. 2008.

T. Asfour, J. Schill, H. Peters, C. Klas, J. Bucker, C. Sander, S. Schulz, A. Kargov, T. Werner, and V. Bartenbach. ARMAR-4: A 63 DOF Torque Controlled Humanoid Robot. In *2013 13th IEEE-RAS International Conference on Humanoid Robots (Humanoids)*, pages 390–396, Oct. 2013.

P. Azad. *Integrating Vision Toolkit*. 2009. URL `http://ivt.sourceforge.net/`.

D. B. Bartolini, R. Cattaneo, G. C. Durelli, M. Maggio, M. D. Santambrogio, and F. Sironi. The Autonomic Operating System Research Project: Achievements and Future Directions. In *Proceedings of the 50th Annual Design Automation Conference*, DAC '13, pages 77:1–77:10, New York, NY, USA, 2013. ACM.

S. Becker, J. Happe, and H. Koziolek. Putting Components into Context: Supporting QoS-Predictions with an explicit Context Model. In R. Reussner, C. Szyperski, and W. Weck, editors, *Proc. 11th International Workshop on Component Oriented Programming (WCOP'06)*, pages 1–6, July 2006.

S. Becker, H. Koziolek, and R. Reussner. The Palladio component model for model-driven performance prediction. *Journal of Systems and Software*, 82(1):3–22, Jan. 2009.

J. Bialkowski, S. Karaman, and E. Frazzoli. Massively parallelizing the RRT and the RRT*. In *2011 IEEE/RSJ International Conference on Intelligent Robots and Systems (IROS)*, pages 3513–3518, Sept. 2011.

A. Bihlmaier, M. Hadlich, and H. Wörn. Advanced ROS Network Introspection (ARNI). In A. Koubaa, editor, *Robot Operating System (ROS)*, number 625 in Studies in Computational Intelligence, pages 651–670. Springer International Publishing, 2016.

S. Browne, J. Dongarra, N. Garner, K. London, and P. Mucci. A Scalable Cross-Platform Infrastructure for Application Performance Tuning Using Hardware Counters. In *Supercomputing, ACM/IEEE 2000 Conference*, pages 42–42, Nov. 2000.

H. Bruyninckx, P. Soetens, and B. Koninckx. The Real-Time Motion Control Core of the Orocos Project. In *IEEE International Conference on Robotics and Automation, 2003. Proceedings. ICRA '03*, volume 2, pages 2766–2771 vol.2, Sept. 2003.

R. Castano, T. Estlin, D. Gaines, A. Castano, C. Chouinard, B. Bornstein, R. Anderson, S. Chien, A. Fukunaga, and M. Judd. Opportunistic Rover Science: Finding and Reacting to Rocks, Clouds and Dust Devils. In *2006 IEEE Aerospace Conference*, page 16 pp., 2006.

P. Charles, C. Grothoff, V. Saraswat, C. Donawa, A. Kielstra, K. Ebcioglu, C. von Praun, and V. Sarkar. X10: An Object-oriented Approach to Non-uniform Cluster Computing. In *Proceedings of the 20th Annual ACM SIGPLAN Conference on Object-oriented Programming, Systems, Languages, and Applications*, OOPSLA '05, pages 519–538, New York, NY, USA, 2005. ACM.

J. Chestnutt, M. Lau, G. Cheung, J. Kuffner, J. Hodgins, and T. Kanade. Footstep Planning for the Honda ASIMO Humanoid. In *Proceedings of the 2005 IEEE International Conference on Robotics and Automation*, pages 629–634, Apr. 2005.

J. A. Colmenares, G. Eads, S. Hofmeyr, S. Bird, M. Moretó, D. Chou, B. Gluzman, E. Roman, D. B. Bartolini, N. Mor, K. Asanović, and J. D. Kubiatowicz. Tessellation: Refactoring the OS Around Explicit Resource Containers with Continuous Adaptation. In *Proceedings of the 50th Annual Design Automation Conference*, DAC '13, pages 76:1–76:10, New York, NY, USA, 2013. ACM.

G. De Giacomo, R. Reiter, and M. Soutchanski. Execution Monitoring of High-Level Robot Programs. In *Proceedings of the 6th International Conference on Principles of Knowledge Representation and Reasoning (KR'98)*, pages 453–465, 1998.

J. Denny, M. Morales, S. Rodriguez, and N. M. Amato. Adapting RRT growth for heterogeneous environments. In *2013 IEEE/RSJ International Conference on Intelligent Robots and Systems (IROS)*, pages 1772–1778, Nov. 2013.

D. Devaurs, T. Siméon, and J. Cortés. Parallelizing RRT on distributed-memory architectures. In *2011 IEEE International Conference on Robotics and Automation (ICRA)*, pages 2261–2266, May 2011.

D. Devaurs, T. Siméon, and J. Cortés. Parallelizing RRT on Large-Scale Distributed-Memory Architectures. *IEEE Transactions on Robotics*, 29 (2):571–579, Apr. 2013.

M. Diftler, J. Mehling, M. Abdallah, N. Radford, L. Bridgwater, A. Sanders, R. Askew, D. Linn, J. Yamokoski, F. Permenter, B. Hargrave, R. Piatt, R. Savely, and R. Ambrose. Robonaut 2 - The First Humanoid Robot in Space. In *2011 IEEE International Conference on Robotics and Automation (ICRA)*, pages 2178–2183, May 2011.

A. Easwaran, M. Anand, and I. Lee. Compositional Analysis Framework Using EDP Resource Models. In *28th IEEE International Real-Time Systems Symposium, 2007. RTSS 2007*, pages 129–138, Dec. 2007.

Elmo Motion Control Ltd. Elmo Servo Drivers, 2016. URL `http://www.elmomc.com/`.

J. Englsberger, A. Werner, C. Ott, B. Henze, M. Roa, G. Garofalo, R. Burger, A. Beyer, O. Eiberger, K. Schmid, and A. Albu-Schaffer. Overview of the torque-controlled humanoid robot TORO. In *2014 14th IEEE-RAS International Conference on Humanoid Robots (Humanoids)*, pages 916–923, Nov. 2014.

O. Faugeras, B. Hotz, H. Mathieu, T. Viéville, Z. Zhang, P. Fua, E. Théron, L. Moll, G. Berry, J. Vuillemin, P. Bertin, and C. Proy. Real time correlation-based stereo: algorithm, implementations and applications. 1993.

J. D. Gammell, S. S. Srinivasa, and T. D. Barfoot. Informed RRT*: Optimal sampling-based path planning focused via direct sampling of an admissible ellipsoidal heuristic. In *2014 IEEE/RSJ International Conference* *on Intelligent Robots and Systems (IROS 2014)*, pages 2997–3004, Sept. 2014.

J. D. Gammell, S. S. Srinivasa, and T. D. Barfoot. Batch Informed Trees (BIT*): Sampling-based optimal planning via the heuristically guided search of implicit random geometric graphs. In *2015 IEEE International Conference on Robotics and Automation (ICRA)*, pages 3067–3074, May 2015.

V. Garousi, L. C. Briand, and Y. Labiche. A UML-based quantitative framework for early prediction of resource usage and load in distributed real-time systems. *Software & Systems Modeling*, 8(2):275–302, Apr. 2009.

GNU Project. The GNU C Library: Process Resource Usage Information, 1997a. URL `http://www.gnu.org/software/libc/manual/html_node/Resource-Usage.html`.

GNU Project. The GNU C Library: Statistics for Memory Allocation, 1997b. URL `http://www.gnu.org/software/libc/manual/html_node/Statistics-of-Malloc.html`.

Google. PerfTools, 2005. URL `https://github.com/gperftools/gperftools`.

S. L. Graham, P. B. Kessler, and M. K. McKusick. Gprof: A Call Graph Execution Profiler. In *Proceedings of the 1982 SIGPLAN Symposium on Compiler Construction*, SIGPLAN '82, pages 120–126, New York, NY, USA, 1982. ACM.

R. Grimm. Ressourcengewahre Bewegungsplanung für humanoide Roboter. Bachelor thesis, Karlsruhe Institute of Technology (KIT), September 2015.

T. Haaß. Situationsabhängige Prädiktion von Rechenressourcen für humanoide Roboter. Bachelor thesis, Karlsruhe Institute of Technology (KIT), April 2014.

F. Hannig, S. Roloff, G. Snelting, J. Teich, and A. Zwinkau. Resource-aware Programming and Simulation of MPSoC Architectures Through Extension of X10. In *Proceedings of the 14th International Workshop on Software and Compilers for Embedded Systems*, SCOPES '11, pages 48–55, New York, NY, USA, 2011. ACM.

J. Happe. *Predicting Software Performance in Symmetric Multi-core and Multiprocessor Environments.* Dissertation, University of Oldenburg, Germany, Aug. 2008.

D. Harel. STATECHARTS: A VISUAL FORMALISM FOR COMPLEX SYSTEMS. *Science of Computer Programming*, 8(3):231–274, June 1987.

D. Harel and A. Naamad. The STATEMATE Semantics of Statecharts. *ACM Trans. Softw. Eng. Methodol.*, 5(4):293–333, Oct. 1996.

J. Heisswolf, S. Friederich, L. Masing, A. Weichslgartner, A. M. Zaib, C. Stein, M. Duden, J. Teich, T. Wild, A. Herkersdorf, and J. Becker. A Novel NoC-Architecture for Fault Tolerance and Power Saving. In *Proceedings of the third International Workshop on Multi-Objective Many-Core Design (MOMAC) in conjunction with International Conference on Architecture of Computing Systems (ARCS)*, pages 1–8, Nuremberg, Germany, Apr. 2016. IEEE.

J. Henkel, L. Bauer, M. Hübner, and A. Grudnitsky. i-Core: A run-time adaptive processor for embedded multi-core systems. In *International Conference on Engineering of Reconfigurable Systems and Algorithms (ERSA)*, 2011.

G. Hirzinger and B. Bauml. Agile Robot Development (aRD): A Pragmatic Approach to Robotic Software. In *2006 IEEE/RSJ International Conference on Intelligent Robots and Systems*, pages 3741 –3748, Oct. 2006.

H. Hoffmann, M. Maggio, M. D. Santambrogio, A. Leva, and A. Agarwal. SEEC: A Framework for Self-aware Management of Multicore Resources. Mar. 2011.

J. Ichnowski and R. Alterovitz. Scalable Multicore Motion Planning Using Lock-Free Concurrency. *IEEE Transactions on Robotics*, 30(5):1123–1136, Oct. 2014.

G. Infantes, F. Ingrand, and M. Ghallab. Learning Behaviors Models for Robot Execution Control. pages 394–397, Ambleside, The English Lake District, U.K., 2006.

S. A. Jacobs, K. Manavi, J. Burgos, J. Denny, S. Thomas, and N. M. Amato. A scalable method for parallelizing sampling-based motion planning algorithms. In *2012 IEEE International Conference on Robotics and Automation (ICRA)*, pages 2529–2536, May 2012.

S. A. Jacobs, N. Stradford, C. Rodriguez, S. Thomas, and N. M. Amato. A scalable distributed RRT for motion planning. In *2013 IEEE International Conference on Robotics and Automation (ICRA)*, pages 5088–5095, May 2013.

L. Jaillet, A. Yershova, S. M. L. Valle, and T. Simeon. Adaptive tuning of the sampling domain for dynamic-domain RRTs. In *2005 IEEE/RSJ International Conference on Intelligent Robots and Systems, 2005. (IROS 2005)*, pages 2851–2856, Aug. 2005.

F. Kanehiro, H. Hirukawa, and S. Kajita. OpenHRP: Open Architecture Humanoid Robotics Platform. *The International Journal of Robotics Research*, 23(2):155–165, Jan. 2004.

K. Kaneko, F. Kanehiro, S. Kajita, H. Hirukawa, T. Kawasaki, M. Hirata, K. Akachi, and T. Isozumi. Humanoid Robot HRP-2. In *2004 IEEE International Conference on Robotics and Automation*, volume 2, pages 1083–1090, Apr. 2004.

K. Kaneko, F. Kanehiro, M. Morisawa, K. Miura, S. Nakaoka, and S. Kajita. Cybernetic Human HRP-4c. In *9th IEEE-RAS International Conference on Humanoid Robots, 2009. Humanoids 2009*, pages 7–14, Dec. 2009.

S. Karaman and E. Frazzoli. Sampling-based algorithms for optimal motion planning. *The International Journal of Robotics Research*, 30(7):846–894, Jan. 2011.

T. Karcher and V. Pankratius. Run-Time Automatic Performance Tuning for Multicore Applications. In E. Jeannot, R. Namyst, and J. Roman, editors, *Euro-Par 2011 Parallel Processing*, number 6852 in Lecture Notes in Computer Science, pages 3–14. Springer Berlin Heidelberg, Aug. 2011.

L. E. Kavraki, P. Svestka, J. C. Latombe, and M. H. Overmars. Probabilistic roadmaps for path planning in high-dimensional configuration spaces. *IEEE Transactions on Robotics and Automation*, 12(4):566–580, Aug. 1996.

Kernel.org. perf, 2010. URL `https://perf.wiki.kernel.org/index.php/Main_Page`.

D. Kissler, F. Hannig, A. Kupriyanov, and J. Teich. A Highly Parameterizable Parallel Processor Array Architecture. pages 105–112. IEEE, Dec. 2006.

M. Klotzbücher and H. Bruyninckx. Coordinating Robotic Tasks and Systems with rFSM Statecharts. *JOSER: Journal of Software Engineering for Robotics*, 3(1):28–56, Dec. 2012.

K. Klues, B. Rhoden, Y. Zhu, A. Waterman, and E. Brewer. Processes and Resource Management in a Scalable Many-core OS. *HotPar10, Berkeley, CA*, 2010.

S. Kobbe, L. Bauer, D. Lohmann, W. Schröder-Preikschat, and J. Henkel. DistRM: Distributed Resource Management for On-chip Many-core Systems. In *Proceedings of the Seventh IEEE/ACM/IFIP International Conference on Hardware/Software Codesign and System Synthesis*, CODES+ISSS '11, pages 119–128, New York, NY, USA, 2011. ACM.

S. Kounev, F. Brosig, N. Huber, and R. Reussner. Towards self-aware performance and resource management in modern service-oriented systems. In *2010 IEEE International Conference on Services Computing (SCC)*, pages 621–624, 2010.

K. Krogmann and R. Reussner. Palladio – Prediction of Performance Properties. In A. Rausch, R. Reussner, R. Mirandola, and F. Plášil, editors, *The Common Component Modeling Example*, number 5153 in Lecture Notes in Computer Science, pages 297–326. Springer Berlin Heidelberg, Jan. 2008.

M. Kröhnert, N. Vahrenkamp, J. Paul, W. Stechele, and T. Asfour. Resource Prediction for Humanoid Robots. *Proceedings of the First Workshop on Resource Awareness and Adaptivity in Multi-Core Computing (Racing 2014)*, pages 22–28, May 2014.

M. Kröhnert, R. Grimm, N. Vahrenkamp, and T. Asfour. Resource-Aware Motion Planning. In *2016 IEEE International Conference on Robotics and Automation (ICRA)*, pages 32–39, May 2016.

J. J. Kuffner and S. M. LaValle. RRT-connect: An efficient approach to single-query path planning. In *IEEE International Conference on Robotics and Automation, 2000. Proceedings. ICRA '00*, volume 2, pages 995–1001 vol.2, 2000.

D. Kulic and E. Croft. Affective State Estimation for Human-Robot Interaction. *IEEE Transactions on Robotics*, 23(5):991–1000, Oct. 2007.

U. Kurup, C. Lebiere, A. Stentz, and M. Hebert. A Hybrid Model for Execution Monitoring in Autonomous Agents. pages 149–154, Carleton University, Ottawa, Canada, 2013.

J. Levon, P. Elie, and M. Johnson. OProfile, 2002. URL `http://oprofile.sourceforge.net/`.

Linux Kernel Documentation. T H E /proc F I L E S Y S T E M. URL `https://www.kernel.org/doc/Documentation/filesystems/proc.txt`.

S. Lohmeier, T. Buschmann, and H. Ulbrich. Humanoid Robot LOLA. In *IEEE International Conference on Robotics and Automation, 2009. ICRA '09*, pages 775–780, May 2009.

R. McDougall, J. Mauro, and B. Gregg. *Solaris(TM) Performance and Tools: DTrace and MDB Techniques for Solaris 10 and OpenSolaris (Solaris Series)*. Prentice Hall PTR, Upper Saddle River, NJ, USA, 2006.

T. Merz, P. Rudol, and M. Wzorek. Control System Framework for Autonomous Robots Based on Extended State Machines. In *2006 International Conference on Autonomic and Autonomous Systems, 2006. ICAS '06*, pages 14–14, July 2006.

G. Metta, P. Fitzpatrick, and L. Natale. YARP: Yet Another Robot Platform. *International Journal on Advanced Robotics Systems*, 3(1):43–48, 2006.

G. Metta, G. Sandini, D. Vernon, L. Natale, and F. Nori. The iCub humanoid robot: an open platform for research in embodied cognition. In *Proceedings of the 8th Workshop on Performance Metrics for Intelligent Systems*, PerMIS '08, pages 50–56, New York, NY, USA, 2008. ACM.

D. Meyer-Delius, C. Plagemann, and W. Burgard. Probabilistic Situation Recognition for Vehicular Traffic Scenarios. In *IEEE International Conference on Robotics and Automation, 2009. ICRA '09*, pages 459–464, May 2009.

G. E. Moore. Cramming more components onto integrated circuits, Reprinted from Electronics, volume 38, number 8, April 19, 1965, pp.114 ff. *IEEE Solid-State Circuits Society Newsletter*, 11(5):33–35, Sept. 2006.

C. Müller-Schloer, H. Schmeck, and T. Ungerer, editors. *Organic Computing — A Paradigm Shift for Complex Systems*. Springer Basel, Basel, 2011.

L. Muscari, L. Seminara, F. Mastrogiovanni, M. Valle, M. Capurro, and G. Cannata. Real-Time Reconstruction of Contact Shapes for Large Area Robot Skin. In *2013 IEEE International Conference on Robotics and Automation (ICRA)*, pages 2360–2366, May 2013.

G. Nelson, A. Saunders, N. Neville, B. Swilling, J. Bondaryk, D. Billings, C. Lee, R. Playter, and M. Raibert. PETMAN: A Humanoid Robot for Testing Chemical Protective Clothing. *Journal of the Robotics Society of Japan*, 30(4):372–377, 2012.

Object Management Group (OMG). UML Profile for Schedulability, Performance, and Time Specification, Version 1.1, 2005. URL `http://www.omg.org/spec/SPTP/1.1/`.

Object Management Group (OMG). UML Profile for MARTE: Modeling and Analysis of Real-Time Embedded Systems, Version 1.1, 2011. URL `http://www.omg.org/spec/MARTE/1.1/`.

Object Management Group (OMG). OMG Unified Modeling Language Version 2.5, 2015. URL `http://www.omg.org/spec/UML/2.5/PDF`.

B. Oechslein, J. Schedel, J. Kleinöder, L. Bauer, J. Henkel, D. Lohmann, and W. Schröder-Preikschat. OctoPOS: A Parallel Operating System for Invasive Computing. In R. McIlroy, J. Sventek, T. Harris, and T. Roscoe, editors, *Proceedings of the International Workshop on Systems for Future Multi-Core Architectures (SFMA)*, volume USB Proceedings of *Sixth*

International ACM/EuroSys European Conference on Computer Systems (EuroSys), pages 9–14, Salzburg, Austria, Apr. 2011. EuroSys.

C. Ott, O. Eiberger, W. Friedl, B. Bauml, U. Hillenbrand, C. Borst, A. Albu-Schaffer, B. Brunner, H. Hirschmuller, S. Kielhofer, R. Konietschke, M. Suppa, T. Wimbock, F. Zacharias, and G. Hirzinger. A Humanoid Two-Arm System for Dexterous Manipulation. In *2006 6th IEEE-RAS International Conference on Humanoid Robots*, pages 276 –283, Dec. 2006.

F. Otto, C. A. Schaefer, M. Dempe, and W. F. Tichy. A Language-Based Tuning Mechanism for Task and Pipeline Parallelism. In P. D'Ambra, M. Guarracino, and D. Talia, editors, *Euro-Par 2010 - Parallel Processing*, number 6272 in Lecture Notes in Computer Science, pages 328–340. Springer Berlin Heidelberg, Aug. 2010.

A. Paraschos, N. I. Spanoudakis, and M. G. Lagoudakis. Model-driven Behavior Specification for Robotic Teams. In *Proceedings of the 11th International Conference on Autonomous Agents and Multiagent Systems - Volume 1*, AAMAS '12, pages 171–178, Richland, SC, 2012. International Foundation for Autonomous Agents and Multiagent Systems.

C. Park, J. Pan, and D. Manocha. Poisson-RRT. In *2014 IEEE International Conference on Robotics and Automation (ICRA)*, pages 4667–4673, May 2014.

J.-K. Park, S.-I. Jin, H.-K. Cho, and Y.-K. Chung. Design and Implementation of Service-Oriented Task Model for Autonomous Service Robot. In *SICE-ICASE, 2006. International Joint Conference*, pages 2623–2628, 2006.

Y.-S. Park, H.-M. Koo, and I.-Y. Ko. A task-based and resource-aware approach to dynamically generate optimal software architecture for intelligent service robots. *Journal of Software: Practice and Experience*, 42(5): 519–541, May 2012.

J. Paul, W. Stechele, M. Kröhnert, T. Asfour, and R. Dillmann. Invasive Computing for Robotic Vision. In *Design Automation Conference (ASP-DAC), 2012 17th Asia and South Pacific*, pages 207–212, Jan. 2012.

O. Pettersson. Execution monitoring in robotics: A survey. *Robotics and Autonomous Systems*, 53(2):73–88, Nov. 2005.

E. Plaku, K. E. Bekris, B. Y. Chen, A. M. Ladd, and L. E. Kavraki. Sampling-based roadmap of trees for parallel motion planning. *IEEE Transactions on Robotics*, 21(4):597–608, Aug. 2005.

R. K. Pujari, T. Wild, A. Herkersdorf, B. Vogel, and J. Henkel. Hardware Assisted Thread Assignment for RISC based MPSoCs in Invasive Computing. pages 106–109. IEEE, Dec. 2011.

M. Quigley, K. Conley, B. Gerkey, J. Faust, T. Foote, J. Leibs, R. Wheeler, and A. Y. Ng. ROS: an open-source Robot Operating System. In *ICRA workshop on open source software*, volume 3, page 5, 2009.

K. Regenstein, T. Kerscher, C. Birkenhofer, T. Asfour, M. Zollner, and R. Dillmann. Universal Controller Module (UCoM) - component of a modular concept in robotic systems. In *2007 IEEE International Symposium on Industrial Electronics*, pages 2089–2094, June 2007.

C. Rodriguez, J. Denny, S. A. Jacobs, S. Thomas, and N. M. Amato. Blind RRT: A probabilistically complete distributed RRT. In *2013 IEEE/RSJ International Conference on Intelligent Robots and Systems (IROS)*, pages 1758–1765, Nov. 2013.

A. Roncone, M. Hoffmann, U. Pattacini, and G. Metta. Automatic kinematic chain calibration using artificial skin: Self-touch in the iCub humanoid robot. In *2014 IEEE International Conference on Robotics and Automation (ICRA)*, pages 2305–2312, May 2014.

S. J. Russell, P. Norvig, J. F. Candy, J. M. Malik, and D. D. Edwards. *Artificial Intelligence: A Modern Approach.* Prentice-Hall, Inc., Upper Saddle River, NJ, USA, 1996.

Y. Sakagami, R. Watanabe, C. Aoyama, S. Matsunaga, N. Higaki, and K. Fujimura. The intelligent ASIMO: System overview and integration. In *2002 IEEE/RSJ International Conference on Intelligent Robots and Systems*, volume 3, pages 2478–2483, 2002.

C. Schlegel, T. Hassler, A. Lotz, and A. Steck. Robotic Software Systems: From Code-Driven to Model-Driven Designs. In *International Conference on Advanced Robotics, 2009. ICAR 2009*, pages 1–8, June 2009.

D. C. Schmidt. The Adaptive Communication Environment: An Object-Oriented Network Programming Toolkit for Developing Communication Software. pages 214–225, 1993.

A. Sharifi, S. Srikantaiah, A. K. Mishra, M. Kandemir, and C. R. Das. METE: Meeting End-to-end QoS in Multicores through System-wide Resource Management. In *Proceedings of the ACM SIGMETRICS Joint International Conference on Measurement and Modeling of Computer Systems*, SIGMETRICS '11, pages 13–24, New York, NY, USA, 2011. ACM.

S. Shende and A. D. Malony. The TAU Parallel Performance System. *International Journal of High Performance Computing Applications*, 20(2): 287–331, 2006.

I. Shin and I. Lee. Periodic Resource Model for Compositional Real-Time Guarantees. In *24th IEEE Real-Time Systems Symposium, 2003. RTSS 2003*, pages 2–13, Dec. 2003.

B. Siciliano and O. Khatib, editors. *Springer handbook of robotics*. Springer, Berlin, 2008.

R. Simmons and D. Apfelbaum. A Task Description Language for Robot Control. In *1998 IEEE/RSJ International Conference on Intelligent Robots and Systems*, volume 3, pages 1931–1937, 1998.

A. Steck and C. Schlegel. Towards Quality of Service and Resource Aware Robotic Systems through Model-Driven Software Development. arXiv e-print 1009.4877, Sept. 2010.

M. Strandberg. Augmenting RRT-planners with local trees. In *2004 IEEE International Conference on Robotics and Automation, 2004. Proceedings. ICRA '04*, volume 4, pages 3258–3262 Vol.4, Apr. 2004.

K. Subramoniam, M. Maheswaran, and M. Toulouse. Towards a Micro-Economic Model for Resource Allocation in Grid Computing Systems. In *Canadian Conference on Electrical and Computer Engineering, 2002. IEEE CCECE 2002*, volume 2, pages 782–785, 2002.

SystemTap Documentation. SystemTap, 2005. URL `https://sourceware.org/systemtap/documentation.html`.

S. Taha, A. Radermacher, S. Gerard, and J. Dekeyser. An Open Framework for Detailed Hardware Modeling. In *International Symposium on Industrial Embedded Systems, 2007. SIES '07*, pages 118–125, July 2007.

J. Teich, J. Henkel, A. Herkersdorf, D. Schmitt-Landsiedel, W. Schröder-Preikschat, and G. Snelting. Invasive Computing: An Overview. In M. Hübner and J. Becker, editors, *Multiprocessor System-on-Chip*, pages 241–268. Springer New York, 2011.

F. Thomas, S. Gérard, J. Delatour, and F. Terrier. Software Real-Time Resource Modeling. In P. E. Villar, editor, *Embedded Systems Specification and Design Languages*, number 10 in Lecture Notes in Electrical Engineering, pages 169–182. Springer Netherlands, Jan. 2008.

U. Thomas, G. Hirzinger, B. Rumpe, C. Schulze, and A. Wortmann. A New Skill Based Robot Programming Language Using UML/P Statecharts. In *2013 IEEE International Conference on Robotics and Automation (ICRA)*, pages 461–466, 2013.

M. Tillmann, T. Karcher, C. Dachsbacher, and W. F. Tichy. Application-independent Autotuning for GPUs. In M. Bader, F. Peters, A. Bode, H.-J. Bungartz, M. Gerndt, and G. R. Joubert, editors, *Parallel Computing: Accelerating Computational Science and Engineering (CSE)*, Advances in Parallel Computing, pages 626–635. IOS Press, 2014.

S. Vacek, T. Gindele, J. Zöllner, and R. Dillmann. Situation classification for cognitive automobiles using case-based reasoning. In *2007 IEEE Intelligent Vehicles Symposium*, pages 704–709, June 2007.

N. Vahrenkamp, M. Wächter, M. Kröhnert, K. Welke, and T. Asfour. The robot software framework ArmarX. *it - Information Technology*, 57(2): 99–111, 2015.

Valgrind Documentation. Massif: a heap profiler, 2004. URL `http://valgrind.org/docs/manual/ms-manual.html`.

M. Wächter, S. Ottenhaus, M. Kröhnert, N. Vahrenkamp, and T. Asfour. The ArmarX Statechart Concept: Graphical Programing of Robot Behavior. *Humanoid Robotics*, page 33, 2016.

J. Weidendorfer. Sequential Performance Analysis with Callgrind and KCachegrind. In M. Resch, R. Keller, V. Himmler, B. Krammer, and A. Schulz, editors, *Tools for High Performance Computing*, pages 93–113. Springer Berlin Heidelberg, 2008.

K. Welke, N. Vahrenkamp, M. Wächter, M. Kroehnert, and T. Asfour. The ArmarX Framework - Supporting high level robot programming through state disclosure. In *GI Annual German Conference on Informatics*, pages 2823–2837, 2013.

A. Yershova, L. Jaillet, T. Simeon, and S. M. LaValle. Dynamic-Domain RRTs: Efficient Exploration by Controlling the Sampling Domain. In *Proceedings of the 2005 IEEE International Conference on Robotics and Automation, 2005. ICRA 2005*, pages 3856–3861, Apr. 2005.

S. Youssefi, S. Denei, F. Mastrogiovanni, and G. Cannata. Skinware 2.0: A real-time middleware for robot skin. *SoftwareX*, 3–4:6–12, Dec. 2015.

ZeroC, Inc. Internet Communication Engine (Ice), 2015. URL `https://zeroc.com/`.